BRUMB P9-DBI-582

3 3045 00055 3996

8701759

	$35.75
621.16	Higgins, Alex.
H53	Boiler room
	questions and
	answers

THE BRUMBACK LIBRARY

OF VAN WERT COUNTY

VAN WERT, OHIO

5/87

BOILER ROOM
QUESTIONS AND ANSWERS

Boiler Room Questions and Answers (2d ed., 1976)
ALEX HIGGINS AND STEPHEN M. ELONKA

Electrical Systems and Equipment for Industry
ARTHUR H. MOORE AND STEPHEN M. ELONKA

Standard Boiler Operators' Questions and Answers
STEPHEN M. ELONKA AND ANTHONY L. KOHAN

Standard Electronics Questions and Answers
STEPHEN M. ELONKA AND JULIAN L. BERNSTEIN
  Volume I, Basic Electronics
  Volume II, Industrial Applications

Standard Industrial Hydraulics Questions and Answers
STEPHEN M. ELONKA AND ORVILLE H. JOHNSON

Standard Instrumentation Questions and Answers
STEPHEN M. ELONKA AND ALONZO R. PARSONS
  Volume I, Measuring Systems
  Volume II, Control Systems

Standard Plant Operators' Manual (2d ed., 1975)
STEPHEN M. ELONKA

Standard Plant Operator's Questions and Answers
STEPHEN M. ELONKA AND JOSEPH F. ROBINSON
  Volume I
  Volume II

Standard Refrigeration and Air Conditioning Questions and Answers
(2d ed., 1973)
STEPHEN M. ELONKA AND QUAID W. MINICH

BOILER ROOM QUESTIONS AND ANSWERS

ALEX HIGGINS

STEPHEN MICHAEL ELONKA

Contributing Editor, Power *magazine; Licensed Chief Marine Steam
Engineer, Oceans, Unlimited Horsepower; Licensed as Regular
Instructor of Vocational High School, New York State;
Member: National Association of Power Engineers (life honorary);
National Institute for the Uniform Licensing of Power
Engineers, Inc., Honorary Chief Engineer; Author:* The Marmaduke Surfaceblow
Story; Standard Plant Operators' Manual; *Co-author:* Standard Plant
Operator's Questions and Answers, *Volumes I and II;*
Standard Refrigeration and Air Conditioning Questions and Answers;
Standard Instrumentation Questions and Answers, *Volumes I and II;*
Standard Electronics Questions and Answers, *Volumes I and II;*
Standard Industrial Hydraulics Questions and Answers;
Standard Boiler Operators' Questions and Answers;
Electrical Systems and Equipment for Industry;
Handbook of Mechanical Packings

Second Edition

McGraw-HILL BOOK COMPANY
New York   St. Louis   San Francisco   Auckland   Bogotá
Düsseldorf   Johannesburg   London   Madrid   Mexico
Montreal   New Delhi   Panama   Paris   São Paulo
Singapore   Sydney   Tokyo   Toronto

8701759

Library of Congress Cataloging in Publication Data

Higgins, Alex.
  Boiler room questions and answers.

  Includes index.
  1. Steam-boilers — Miscellanea.   I.   Elonka, Stephen
Michael, joint author.   II.   Title.
TJ289.H55   1976   621.1′6′076   76-5499
ISBN   0-07-028754-6

Copyright © 1976, 1945 by McGraw-Hill, Inc. All rights reserved.
Printed in the United States of America. No part of this
publication may be reproduced, stored in a retrieval system, or
transmitted, in any form or by any means, electronic, mechanical,
photocopying, recording, or otherwise, without the prior written
permission of the publisher.

8 9 10 11   BKP BKP   8 9 8 7 6 5 4

The editors for this book were Tyler G. Hicks and Carolyn Nagy,
the designer was Naomi Auerbach, and the production supervisor was George Oechsner.
It was set in Baskerville by Progressive Typographers.
It was printed and bound by The Book Press.

Dedicated to
the boiler operator, licensing examiner, inspector, and all others involved with boilers who must keep power and process services flowing efficiently, safely, and pollution free

H   RYLAND LIBRARY
MANCHESTER

CONTENTS

PREFACE

Thousands in the United States and Canada have been helped to qualify for higher-paying jobs by the first edition of this popular book, a solid basic review of typical examination questions on boiler room operation, written in clear and concise language that is easily understood by all practical power plant operators.

Though the primary purpose of this second edition is to assist those who are studying for licenses (certificates) as firemen and operating engineers, it also provides an excellent basic reference to help the plant operator generate steam efficiently and solve problems that may come up in a plant.

To this second edition we have added a chapter on the latest job opportunities for licensed firemen and stationary engineers in the United States and in Canada, where, incidentally, the requirements are, on the whole, more uniform and stringent than in the United States. And because many stationary engineers end up operating marine plants, and many marine engineers settle down in shoreside plants, we include the U.S. Coast Guard license requirements for merchant marine licenses.

The fuel oil crisis today is again making coal-fired boilers a necessity in many localities. Thus we retain the basic information on coal-burning designs. But new material has been included on burning coal, fuel oil, and gas more efficiently. And because of strict air pollution laws today, information on ways to maintain a "clean stack" and on *safe* operating methods are included. To assist the reader in acquiring a well-rounded technical background, suggested reading material has been added at the end of each chapter.

"Why do boilers explode, although fully protected by safety controls and interlocks?" we asked leading boiler insurance inspectors. The answers in Chap. 8 should not only be known by every owner of boilers, but should be posted in every boiler room. This edition also contains basic information on automatic combustion controls used today.

Parts of this volume have appeared serially in *Power* magazine, where the editors have been extremely helpful. Thanks are also due to the many manufacturers of standard engineering equipment who furnished illustrations and data, and also to various machinery insurance firms.

Our deep thanks go to the many license examiners in the United States and Canada for furnishing sample questions and making helpful suggestions to enable the applicant to obtain a passing grade on his examination.

Alex Higgins
Stephen Michael Elonka

BOILER ROOM
QUESTIONS AND ANSWERS

1

JOB OPPORTUNITIES
IN UNITED STATES
AND CANADA

Today, as never before, the world depends on energy, most of which is produced by burning fuel (coal, oil, gas) in boilers to produce steam, which spins turbine-generators to generate electricity, which turns the wheels of industry and provides electric lighting.

Steam today is also generated in nuclear reactors, in which the reactor is the steam generator (boiler), the remainder of the plant being a conventional steam plant. In fact, at least one state (Massachusetts) now requires nuclear plant operators to have a steam license in addition to the Atomic Energy Commission's license. And at least one municipality, Fulton County, Georgia, now issue stationary engineer's licenses to females.

At this writing (1976) there are 30 nuclear plants in operation in the United States, and 1,000 are projected to be supplying energy by the year 2000.

Operating steam generating plants offers a bright future for any young person who is mechanically inclined and is ambitious enough to qualify for the job classifications listed in this chapter, in which we show the job requirements in the United States and Canada. And because steam power plants are basically similar, whether ashore or afloat, we also include merchant marine license opportunities in the United States. (Indeed, many of the AEC-licensed operators received their nuclear training aboard nuclear submarines in the United States Navy.)

From Tables 1-1, 1-2, and 1-3 the reader can quickly see what must be studied to qualify for any job from fireman right up to chief engineer, and also the capacity of the plant each license entitles the holder to operate.

Understanding boilers and their operation is the *first* step for the embryo stationary or marine engineer, who may not have a technical education. Thus this chapter provides basic information guiding the reader to an interesting career that often leads to travel and higher pay.

TABLE 1-1 United States Stationary Engineer's License Requirements

States and cities	Class license	Education, experience, and remarks	Plant capacity for class license
Alabama	No state license		
Mobile: Board of Engineers Examiners, City Hall, Mobile, 36602	1. Fireman (special permit)	Age 21, 12 months practical experience	One boiler only, up to 50 psi
	2. Wide-open engineer	Same as above	Unlimited
Alaska Dept. of Labor, Juneau, 99801	1. Fireman apprentice	Depends on examiner's opinion	Under supervision of licensed operator
	2. Boiler operator, third class	Same as above	Up to 3,500 lb/hr
	3. Boiler operator, second class	Same as above	Up to 100,000 lb/hr
	4. Boiler operator, first class	Same as above	Unlimited
Arizona	No state license		
Arkansas Boiler Inspection Div., Capitol Hill Bldg., Little Rock, 72203	1. Boiler operator, low grade	6 months on-the-job training	Up to 15 psi steam, or 30 psi water for hot-water boiler
	2. Boiler operator, high grade	Holder of low-grade license	Unlimited

3

TABLE 1-1 United States Stationary Engineer's License Requirements (Continued)

States and cities	Class license	Education, experience, and remarks	Plant capacity for class license
California	No state license		
Los Angeles: Dept. Buildings & Safety, City Hall, Los Angeles, 90012	1. Auxiliary engine operator	Age 19; 3 months as oiler, wiper; or 90-day training course, 60 days of which is operating, oiling, wiping engines or turbines	Work under licensed operating engineer
	2. Steam engine	Age 21; 1 year as steam engine operator; or 3 years as oiler, wiper, or auxiliary engine operator	50 hp or more
	3. Steam engineer	Age 21, 3 years as fireman, water tender, or assistant engineer	500 hp and up
	4. Steam engineer, unlimited	Age 21, 3 years as fireman, water tender, or assistant engineer; or 1 year in charge of steam plant	Unlimited
	5. Boiler operator	Age 21, no experience necessary	Up to 35 boiler hp
San Jose: Bureau of Fire Prevention, 476 Park Ave., San Jose, 95110	1. Special boiler operator	Age 18, sufficient experience to operate a specific boiler	Up to 30 boiler hp, but at only one station
	2. Fireman	Age 18, same as above	Work under licensed engineer
	3. Second-grade steam engineer	Age 21, 3 years operating and maintaining boilers, 2 years of which with high-pressure type	Up to 500 boiler hp unless under first-grade engineer

License	Requirements	Scope
Colorado	No state license	
Denver: Bldg. Inspection Dept., City & County Bldg., Denver, 80203		
1. Boiler operator, class B	Age 21, 3 years boiler operation or equivalent, 2 years high school	15 to 100 psi and 10 to 100 boiler hp
2. Boiler operator, class A	Age 21, 3 years boiler operation or equivalent, high school graduate	Over 15 psi and 10 boiler hp for steam boiler, 250°F in entire system for hot-water boiler
3. Stationary engineer, class A	Age 21, 3 years operating, maintaining boilers, refrigeration machinery; high school graduate	Unlimited steam plant, including refrigeration
4. First-grade steam engineer	Age 25, 5 years in charge of boilers, of which 3 years are with high-pressure type	Unlimited steam power plant
Pueblo: City Engineer, City Hall, Pueblo, 80103		
1. Boiler tender, class D	Age 21, experience not specified	Under supervision of licensed engineer
2. Boiler operator, class C	Age 21, experience not specified	Low-pressure boiler over 300-sq-ft heating surface
3. Operating engineer, class B	Age 21, 2 years of operation	Any boiler or steam engine except automatically fired boilers
4. Chief engineer, class A	Age 21, 5 years operation of boilers and associated machinery	Unlimited steam plant
Connecticut	No state license	

TABLE 1-1 United States Stationary Engineer's License Requirements *(Continued)*

States and cities	Class license	Education, experience, and remarks	Plant capacity for class license
Bridgeport: Power Engineers Board of Examiners, City Hall, Bridgeport, 06600	1. Low-pressure boiler operator	Age 21; 6 months as boiler or water tender, or assistant to licensed engineer, boiler or water tender; or 6 months at technical school; or 1 year operating low-pressure boilers	Up to 15 psi
	2. Boiler operator or water tender	Age 21; 1 year as boiler or water tender, oiler, or assistant to licensed operator or boiler or water tender of plants over 15 psi or 25 boiler hp; or 1 year of technical school and 6 months operating steam boilers	Over 25 boiler hp and 15 psi
	3. Power engineer	Age 21; 4 years as boilermaker or machinist and 1 year in steam boiler plants; or 2 years engineering school and 1 year in steam power plant; or holder of state or U.S. operating engineer license; or 3 years boiler or water tending and work with boilers and steam engines	Unlimited capacity
Delaware	No state license		
Wilmington: Board of Examining Engineers, Public Bldg., Wilmington, 19801	1. Fireman	Age 18, 6 months assisting fireman or engineer with boilers over 15 psi	Boilers of over 15 psi
	2. Engineer	Age 21, 1 year operating, maintaining steam power plants	Unlimited

Agency	License	Requirements	Scope
District of Columbia Dept. of Occupations & Professions, 614 H St. NW, Washington, D.C. 20001	1. Steam engineer, class 6	Age 21, 6 months with steam boilers	Up to 200 boiler hp, 15 psi
	2. Steam engineer, class 5	Age 21, 1 year firing steam boilers	Heating plants under 15 psi
	3. Steam engineer, class 4	Age 21, 1 year with portable boilers or steam engines	Any portable boiler and auxiliaries
	4. Steam engineer, class 3	Age 21, 2 years in high-pressure plants as assistant engineer, fireman, or oiler – 1 year allowed for mechanical engineer degree	In charge of plants up to 250 boiler hp
	5. Steam engineer, class 2	Age 21, same experience as class 3 plus 1 year in charge of class 3 steam plant	In charge of plant up to 750 boiler hp
	6. Steam engineer, class 1	Age 21, same experience as class 3 plus 2 years as class 2 and assisting chief engineer in charge of class 1 plant	Unlimited
Florida	No state license		
Tampa: Boiler Bureau, 301 N. Florida Ave., Tampa, 33602	1. Fireman, low-pressure	Age 21, no experience if passing mark on test	Up to 15 boiler hp
	2. Fireman, high-pressure	Age 21, 6 months firing or assisting first- or second-class engineer of up to 25 boiler hp, 15 psi plant	Up to 25 boiler hp, 15 psi

TABLE 1-1 United States Stationary Engineer's License Requirements (Continued)

States and cities	Class license	Education, experience, and remarks	Plant capacity for class license
	3. Engineer, third class	Age 21, 1 year with steam boilers and steam/internal combustion engines	In charge of portable boilers and engines
	4. Engineer, second class	Age 21; 2 years as oiler, fireman, or assistant to licensed first-class engineer; or technical school grad and 1 year as oiler, fireman, or assistant to licensed first-class engineer	In charge of up to 175-boiler-hp plant
	5. Engineer, first class, grade B	Not specified	Not specified
	6. Engineer, first class, grade A	Age 21; (a) 3 years as oiler or assistant to licensed first-class engineer; or (b) 3 years firing or assistant engineer of steamboat or steam locomotive with 1 year as assistant to first-class engineer; or (c) 2 years machinist or boilermaker plus apprentice time and 2 years as assistant to first-class engineer; or (d) mechanical engineer degree and 1 year under licensed first-class engineer; or (e) 2 years as second-class engineer while holding second-class engineer's license; or (f) holder of first- or second-grade license issued by U.S. Government or U.S. Coast Guard	Unlimited

State	Agency	License	Requirements	Scope
Georgia	No state license			
	Fulton County: Inspections & Licenses Dept., 165 Central Ave. SW, Atlanta, 30303	1. Fireman	Age 18; male or female; 1 year in boiler room; deductible time for mechanical, technical training	Pressure vessels, boilers and engines, and steam hot-water systems; class license for plant not specified
		2. Class 4 engineer	Age 18; male or female; 1 year in power plant; time allowed for machinist, boiler maker, inspector, and technical training	Same as above
		3. Class 3 engineer	Age 21, male or female, 30 months as fireman or class 4 engineer, up to 2 years' time allowed for above	Same as above
		4. Class 2 engineer	Age 21, male or female, 4 years as class 3 engineer, or up to 2 years' time allowed for above	Same as above
		5. Class 1 engineer	Age 21, male or female, 5 years as class 2 engineer, or up to 2 years' time allowed for above	Same as above
Guam	No license			
Hawaii	No state license			
Idaho	No state license			
Illinois	No state license			

TABLE 1-1 United States Stationary Engineer's License Requirements (*Continued*)

States and cities	Class license	Education, experience, and remarks	Plant capacity for class license
Chicago: Boiler & Pressure Vessel Inspection, 320 N. Clark St., Chicago, 60602	1. Boiler tender	Age 21, proof of familiarity with boiler operation and construction	All boilers over 10 psi
	2. Water tender	Same as above	Same as above
	3. Engineer	Age 21; 2 years as machinist or engineer related to boilers, steam engines	Same as above
Decatur: Board of Examiners, Steam Engineers, 62500	1. Second-grade engineer	Age 21, 6 months with boilers, steam equipment	Boilers over 15 psi, up to 75 boiler hp
	2. First-grade engineer	Age 21, 2 years in charge of steam boilers, engines	Unlimited
East St. Louis: Boiler & Elevator Inspector, City Hall, East St. Louis, 66201	1. Engineer	3 years under licensed engineer in charge of power plant	All boilers over 10 boiler hp and 15 psi
Elgin: City Hall, Elgin, 60120	First-class engineer	Age 21, 1 year with steam-generating equipment	All boilers over 15 psi

Evanston: Building Dept., Municipal Bldg., Evanston, 60200	1. Water tender	Age 21, must prove familiarity with steam-generating equipment	All boilers, pressure vessels above 15 psi
	2. Engineer	Age 21, engineer or machinist with 2 years of managing, operating, or constructing boilers, steam engines, or ice machines	Unlimited
Peoria: City Boiler Inspector, City Hall, Peoria, 61602	1. Boiler tender	Age 21, 3 years in boiler room	All boilers over 15 psi
	2. Second-class engineer	Same as above	Same as above
	3. First-class engineer	Same as above	Same as above
Indiana	No state license		
Hammond: Board of Examiners, City Hall, Hammond, 46300	1. Water tender	Age 21	Any steam boilers
	2. Engineer	Age 21, 2 years in charge of steam boilers and engines— 1 year if 2 years as machinist in steam-engine works or 1 year in steam plant if degreed engineer	Same as above
Terre Haute: Board of Examiners, City Hall, Terre Haute, 47800	1. Third-class license	Age 21, 2 years firing steam boilers, or 1 year in charge of steam engines, turbines, boilers	Up to 50 boiler hp
	2. Second-class license	Same as above	Up to 200 boiler hp
	3. First-class license	Age 21, 5 years operating steam plants	Unlimited

TABLE 1-1 United States Stationary Engineer's License Requirements *(Continued)*

States and cities	Class license	Education, experience, and remarks	Plant capacity for class license
Iowa	No state license		
Des Moines: Dept. Building Inspectors, City Hall, Des Moines, 50309	1. Fireman, second class	Age 21, 1 year helping fireman in boiler plant	Heating plant up to 50 boiler hp
	2. Fireman, first class	Age 21, 2 years as fireman or fireman's helper in boiler plant	Heating plant up to 75 boiler hp
	3. Third-class engineer	Same as above	Steam plant up to 125 boiler hp
	4. Second-class engineer	Age 21, 3 years in steam plant and knowledge of refrigeration, heating, ventilation and electrical apparatus	Steam plant up to 200 boiler hp
	5. First-class engineer	Age 21, 5 years in steam engineering or refrigeration plants, with experience in operating such plants	Unlimited
Sioux City: City Hall, Sioux City, 51102	1. Third-grade engineer	3 years in steam plants	Up to 15 psi
	2. Second-grade engineer	Same as above	Up to 40 psi
	3. First-grade engineer	5 years in steam plants	Up to 125 psi
Kansas	No state license		
Kentucky	No state license		

	Requirements	Scope
Covington: Examiner of Engineers, City Hall, Covington, 41000		
1. Second-class engineer	Not specified	All boilers over 10 psi
2. First-class engineer	Not specified	Same as above
Louisiana		
No state license		
New Orleans: Mechanical Inspection Section, City Hall, New Orleans, 70112		
1. Hoisting & portable engineer	Age 18, 2 years with steam, internal combustion, or electrical equipment	Unlimited, portable boilers
2. Special operator	Age 18, 2 years operating steam boilers of more than 30 gal water and up to 240-sq-ft heating surface, of up to 15 psi	Not specified
3. Third-class operating engineer	Age 18; 2 years operating steam engines or boilers from 20 to 50 boiler hp; or at least 2 years apprentice engineer with steam engines or boilers of first-, second-, or third-class hp rating; or hold mechanical engineer degree	Not specified
4. Second-class operating engineer	Age 18, 3 years operating steam engines or boilers from 51 to 150 boiler hp, or 6 months in steam plants while holding mechanical engineer degree	Not specified
5. First-class operating engineer	Age 18, 4 years as engineer operating equipment 150 boiler hp and over, or 1 year in steam plants while holding mechanical engineer degree	Unlimited

TABLE 1-1 United States Stationary Engineer's License Requirements *(Continued)*

States and cities	Class license	Education, experience, and remarks	Plant capacity for class license
Maine Boiler Rules & Regulations, State Office Bldg. Annex, Augusta, 04330	1. Boiler operator, low-pressure, heating	Any engineer or fireman who has operated in steam plants for 1 year	Heating plant with steam boilers up to 15 psi, or hot-water boilers up to 160 psi or 250°F, or both
	2. Boiler operator	6 months experience in steam power plants	Heating plant up to 20,000 lb/hr, or operate in any plant under engineer licensed for that plant
	3. Fourth-class engineer	High school graduate or equivalent and 1 year operating under licensed engineer in charge of plant	50,000 lb/hr plant, or operate in any plant under engineer licensed for plant
	4. Third-class engineer	1 year operating on fourth-class license	100,000 lb/hr, or operate in plant under engineer licensed for plant
	5. Second-class engineer	2 years operating on third-class license	200,000 lb/hr, or operate in any plant under engineer licensed for plant
	6. First-class engineer	2 years operating on second-class license	Unlimited capacity
Maryland Department of Licensing & Regula- tions, 203 E. Balti- more St., Baltimore, 21202	1. Fourth-grade engineer	Age 21, 2 years with steam boilers and engines	Hoisting or portable boilers
	2. Third-grade engineer	Same as above	Up to 30 boiler hp
	3. Second-grade engineer	Same as above	Up to 500 boiler hp
	4. First-grade engineer	Same as above	Unlimited

Massachusetts Board of Boiler Rules, 1010 Commonwealth Ave., Boston, 02215		
1. Second-class fireman	Not specified	Any capacity under third-class fireman or engineer
2. First-class fireman	1 year with boilers while holding second-class fireman license, or 1 year as engineer or fireman operating boilers	Boilers with safety valves set not over 25 psi, or operate any boilers under engineer or higher-grade fireman
3. Extra first-class fireman	Same as above	Unlimited boiler capacity
4. Portable-class engineer	Not specified	Any portable boilers and engines
5. Special engineer	Not specified	Up to 150-hp engine in specific plant
6. Fourth-class engineer	Not specified	Portable boilers and steam engines
7. Third-class engineer	1½ years on steam engineer or fireman license, or 1 year as fireman while holding fourth-class engineer license and working under first-class licensed fireman	Up to 150 boiler hp of not over 50 hp each, or second-class plant under licensed engineer
8. Second-class engineer	2 years in charge of at least one engine of over 50 hp; or 1 year as engineer, assistant engineer, or fireman while holding third-class engineer license; or 2 years working on special license; or 1 year in steam plants after 3 years machinist apprenticeship or boilermaker in stationary, marine, or boiler works; or 1 year in steam plants while holding mechanical or chemical engineer degree	Up to 150-boiler hp (each), or operate in first-class plant under license engineer

TABLE 1-1 United States Stationary Engineer's License Requirements (Continued)

States and cities	Class license	Education, experience, and remarks	Plant capacity for class license
	9. First-class engineer	3 years in charge of plant with one 150-hp engine, or 1½ years in second-class or first-class plant while working on second-class license	Unlimited steam plant
	10. Assistant nuclear power plant operator	1 year in steam plant as fireman, water tender, control room assistant to engineer in charge of 8-hour day shift. (Massachusetts is the only state thus far requiring steam license in nuclear plant.)	Any nuclear plant; operate auxiliaries, control room and/or related systems under supervision of nuclear power plant operating engineer, or nuclear power plant senior supervising engineer. NOTE: This license does not entitle holder to operate plant using fossil fuel as major source of heat energy.
	11. Nuclear power plant operating engineer	1½ years in steam or nuclear-steam power plant on second-class engineer license or assistant nuclear steam license. 1 year of nuclear plant engineering school equivalent to assistant nuclear operator license, but applicant must complete 6 months operating as assistant nuclear-steam power plant operator, or have B.S. in engineering and 1 year in steam plant.	Any nuclear plant; shift supervisor in charge of nuclear power plant, its auxiliaries, prime movers, related control systems
	12. Nuclear power plant senior supervising engineer	2 years in steam plant on Massachusetts first-class engineer or nuclear power plant engineer license. 1 year	Any nuclear plant; in full charge of entire nuclear power plant, prime movers, auxiliaries, control systems

of nuclear power plant engineering courses equivalent to nuclear power plant operating engineer experience.

as the designated shift supervisor

Michigan

No state license

Authority	License	Requirements	Scope
Bay City: Boiler Inspector, City Hall, Bay City, 48706	1. Fourth-class engineer	Age 21	In charge of any heating boilers
	2. Third-class engineer	Age 21	Up to 20 boiler hp
	3. Second-class engineer	Age 21	Up to 50 boiler hp
	4. Fourth-class engineer	Age 21	Unlimited boiler hp
Dearborn: Bldg. & Safety Division, City Hall, Dearborn, 48126	1. Boiler operator	Age 21, 2 years in boiler room	NOTE: Boilers over 15 psi, all commercial refrigerating plants over 5 tons, all ammonia systems, and compressed air tanks and pressure vessels require licensed operators, but class license not specified
	2. Fourth-class engineer	Age 21, 2 years in steam plant	
	3. Third-class engineer	Same as above	
	4. Second-class engineer	Age 21, 3 years in steam plant	
	5. First-class engineer	Age 21, 5 years in steam plant or 2 years in steam plant if degreed engineer	
Detroit: Dept. of Bldg. & Safety Engineering, City-County Bldg., Detroit, 48426	1. Miniature boiler operator	Age 18, 1 month operating high-pressure boilers	Up to 16-in. inside-diameter boiler shell, 5-cu-ft gross volume of furnace and 100 psi gage
	2. Low-pressure boiler operator	Age 18, 1 year with high-pressure boilers and steam prime movers	Low-pressure plants up to 5,000-sq-ft boiler heating surface

TABLE 1-1 United States Stationary Engineer's License Requirements (*Continued*)

States and cities	Class license	Education, experience, and remarks	Plant capacity for class license
	3. High-pressure boiler operator	Age 18, 2 years with low- or high-pressure boilers, or 1 year on low-pressure license	Boiler plants of all pressures, but not over 4,000-sq-ft heating surface, and engine-turbine up to 10 hp
	4. Portable steam equipment operator	Age 19, 1 year with high-pressure boilers or steam prime movers	Steam locomotives and portable boilers to 2,000-sq-ft heating surface: portable steam equipment to 150 hp
	5. Third-class stationary engineer	Age 20; 1 year on high-pressure boiler operator license; or 1 year on low-pressure boiler license and 1 year with high-pressure boiler; or 1 year in high-pressure boiler plant of over 4,000-sq-ft heating surface on high-pressure license; or 3 years in high-pressure boiler plant of over 4,000-sq-ft heating surface; or 1 year in high-pressure boiler plant of over 4,000 sq ft plus 3 years with steam prime movers of over 10 hp; or 3 years in high-pressure plant up to 4,000-sq-ft heating surface	Boilers up to 7,500-sq-ft heating surface and up to 100 steam engine-turbine hp
	6. Second-class stationary engineer	Age 21; 1 year on third-class stationary engineer license; or M.E. or E.E. degree and 1 year in steam-electric power plant; or 4 years in high-pressure boiler plant of 7,500-sq-ft heating surface; or 1 year in high-pressure boiler plant of over	Boiler plants up to 20,000-sq-ft heating surface and steam engine-turbine up to 200 hp

	License	Experience	Capacity
	7. First-class stationary engineer	7,500 sq ft plus 4 years with steam prime movers over 100 hp; or 1 year in high-pressure boiler plant of over 7,500 sq ft plus 4 years in high-pressure boiler plant of over 4,000-sq-ft heating surface	Unlimited capacity
Grand Rapids: Inspection Services, City-County Bldg., Grand Rapids, 49502		Age 22, 6 years in high-pressure boiler plant of over 20,000-sq-ft heating surface plus 6 years with prime movers of over 200 hp, or 2 years with high-pressure boilers of over 20,000 sq ft plus 6 years with high-pressure boilers of over 7,500 sq ft	
	1. Boiler operator	1 year with high-pressure boiler	All boilers of 10 boiler hp and 15 psi must have licensed operator. NOTE: No plant capacity specified.
	2. Boiler engineer	4 years with high-pressure boiler	Same as above
Saginaw: Board of Examiners, Stationary Engineers, City Hall, Saginaw, 48602	1. Fireman, limited-horsepower	Age 21, grammar school education and 2 years firing boilers	Same as above
	2. Fireman, unlimited-horsepower	Same as above	Same as above
	3. Engineer, low-pressure	Same as above	Same as above
	4. Engineer, high-pressure	Same as above	Same as above

TABLE 1-1 United States Stationary Engineer's License Requirements (Continued)

States and cities	Class license	Education, experience, and remarks	Plant capacity for class license
Minnesota Division of Boiler Inspection, State Office Bldg., St. Paul, 55155	1. Third-class engineer	Age 18, must satisfy inspector that applicant can safely operate low-pressure boiler of up to 30 boiler hp	All boilers require licensed operators except those in private residence or apartment houses under four families.
	2. Third-grade engineer	Age 18, 6 months with steam boilers up to 30 boiler hp	
	3. Second-class engineer, grade C	Age 19, 1 year with low-pressure boilers of up to 100 boiler hp	Up to 100 boiler hp
	4. Second-class engineer, grade B	Age 20, 1 year with steam boilers of up to 100 boiler hp	Up to 100 boiler hp
	5. Second-class engineer, grade A	Age 21, 1 year with all classes steam boilers, steam engines, or turbines	Up to 100 hp steam boiler, engine, or turbine
	6. First-class engineer, grade C	Age 21, 3 years with all classes low-pressure steam plants	Up to 300 boiler hp, low-pressure
	7. First-class engineer, grade B	Age 21, 3 years with all classes steam boiler plants	Up to 300-boiler-hp plants
	8. First-class engineer, grade A	Age 21, 3 years with all classes steam boilers, engines or turbines	Up to 300-boiler-hp plants, engines, turbines
	9. Chief engineer, grade C	Age 21, 5 years with all classes low-pressure steam plants	All classes low-pressure boilers

Location / Authority	License	Requirements	Scope
	10. Chief engineer, grade B	Age 21, 5 years with all classes steam boilers	All classes steam boilers
	11. Chief engineer, grade A	Age 21, 5 years with all classes steam boilers, engines or turbines	Unlimited steam plant
Mississippi	No state license		
Missouri	No state license		
Kansas City: Buildings & Inspectors Div., City Hall, Kansas City, 64106	1. Portable engineer	Age 21, 2 years with portable boilers and equipment	Portable boilers
	2. Fireman, class A	Age 21, must satisfy examiner that applicant can safely operate class B boiler equipment	Boilers over 15 psi and up to 126 psi
	3. Engineer, class B	Age 21, 3 years with fuel-burning equipment, or 2 years with fuel-burning equipment while holding M.E. degree from school accredited by the state of Missouri	In charge of class B plant (not specified)
	4. Engineer, class A	Age 21, 5 years with all kinds of power plant equipment, know ASME codes for power plant equipment, or M.E. degree from state of Missouri approved school and 3 years in power plants	Unlimited steam plants
St. Joseph: Dept. of Public Works, City Hall, St. Joseph, 64501	1. Special permit	Prove to examiner competence and experience to have charge and operate power, heating, and refrigeration equipment	Only specified equipment under first-class licensed operator

TABLE 1-1 United States Stationary Engineer's License Requirements (*Continued*)

States and cities	Class license	Education, experience, and remarks	Plant capacity for class license
	2. Heating engineer	Same as above	All boilers of 15 psi or less for heating purposes only
	3. Second-class engineer	Same as above	All boilers of 15 psi or more
	4. First-class engineer	3 years practical experience in boiler room or power plant, or equivalent	Unlimited steam boilers, pressure vessels, refrigeration and air-conditioning units
St. Louis: Dept. Public Safety, City Hall, St. Louis, 63103	1. Boiler operator	Must prove familiarity with operation and safety	15 psi to 125 psi of up to 40-sq-ft heating surface
	2. Class 3 stationary engineer	Age 19, must demonstrate skill, ability, and competence to operate boilers	15 psi to 125 psi of 40-sq-ft to 500-sq-ft heating surface
	3. Class 2 stationary engineer	Age 19, 1 year with boilers under class 1 or class 2 engineer; or 1 year maintaining steam boilers or engines; or registered professional engineer; or engineer in training for P.E.	15 to 300 psi boilers up to 2,000-sq-ft heating surface
	4. Class 1 stationary engineer	Age 21, 2 years on class 2 license, or P.E. in training and employed for 1 year in engineering or research of power-generating plant	Any boilers over 15 psi and associated power plant components, or direct-fired gas turbines of over 500 hp

	License	Experience	Scope
Montana Dept. of Labor & Industry, 815 Front St., Helena, 59601	1. Traction engineer	Age 18, 6 months as assisting traction engineer	Any steam traction unit
	2. Low-pressure engineer	Age 18, 3 months operating low-pressure boiler	Up to 15 psi steam, 50 psi hot-water boiler
	3. Third-class engineer	Age 18, 6 months with boilers in this classification	Up to 100 psi steam, 150 psi hot-water boiler
	4. Second-class engineer	Age 20, 2 years with boilers and steam-driven machinery in this classification under second- or first-class engineer, or hold third-class license and 1 year with steam units under second- or first-class engineer	Up to 250 psi steam, 375 psi hot-water boiler
	5. First-class engineer	Age 21; 3 years with steam boilers and units under first-class engineer; or hold second-class license and 1 year with steam boilers under first-class engineer; or hold third-class license and 2 years under first-class engineer	Unlimited steam power plants
Nebraska	No state license		
Lincoln: Board of Examiners, City-County Bldg., Lincoln, 68508	1. Fireman	Age 21, 1 year operating steam boilers	All boilers of 15 psi and over require licensed operator, but class license not specified
	2. Third-grade engineer	Age 21, 1 year operating steam boilers	
	3. Second-grade engineer	Age 21, 3 years operating steam boilers	

23

TABLE 1-1 United States Stationary Engineer's License Requirements *(Continued)*

States and cities	Class license	Education, experience, and remarks	Plant capacity for class license
	4. First-grade engineer	Age 21, 5 years operating steam boilers	
Omaha: Dept. of Public Safety, City Hall, Omaha, 68102	1. Fireman	Age 21, 1 year with steam boilers under licensed operator	Up to 15 psi and heating surface up to 750 sq ft
	2. Limited stationary engineer	Age 21, 1 year with steam boilers under licensed operator of like or higher grade	Up to 100-hp steam prime movers
	3. Third-grade engineer	Same as above	Any steam boiler plant but no prime mover
	4. Second-grade engineer	Age 21, 3 years with steam boilers under licensed engineer of like or higher grade	Any steam power plant up to 100 boiler hp
	5. First-grade engineer	Age 21, 5 years with steam power and/or heating plants	Any steam-power, refrigeration, or compressor plant
Nevada	No state license		
New Hampshire	No state license		
New Jersey Trenton: Mechanical Inspection Bureau, Trenton, 08625	1. Boiler fireman, low-pressure	Age 17, 3 months fireman helper or coal passer	(a) Steam or hot-water boilers over 15 psi and 6 boiler hp; (b) steam or hot-water heating plant above 499-sq-ft heating surface, 1,000 kW, or 4,000,000 Btu; (c) prime movers over 6 hp; (d) 6-ton system with flammable or toxic refrigerant

License	Requirements	Privileges
2. Boiler fireman-in-charge, low pressure	Age 17, 3 months in boiler room under licensed operator	Act as chief of 500 boiler hp plant or assume shift under chief engineer of up to 1,000 boiler hp plant
3. Boiler fireman-in-charge, high pressure	Age 17, 6 weeks training by licensed operator; 6 months as low-pressure fireman reduces 6 weeks to 30 days	Same as above
4. Third-grade engineer (blue seal)	Age 18, 6 months while holding fireman-in-charge license in third-grade engineer plant, or as assistant in such plant	Act as chief engineer of up to 1,000 boiler hp, 100 engine hp, or 65 tons refrigeration, or operate in any plant under chief engineer
5. Second-grade engineer (red seal)	Age 18, 1 year on third-grade license, or equivalent experience	Act as chief engineer of up to 3,000 boiler hp, 500 engine hp, or 300 tons refrigeration, or operate in any plant under chief engineer
6. First-grade engineer (gold seal)	Age 18, 1 year on second-grade license as supervising or chief engineer, or 2 years as operating engineer under first-grade engineer, or equivalent experience	Act as chief engineer of any steam generating capacity
New Mexico	No state license	
New York	No state license	
Buffalo: Division Fuel Devices, City Hall, Buffalo, 14202		
1. Second-class engineer	Age 21, 3 years as fireman, oiler, or helper on repairs of boilers and engines	30 to 100 boiler hp and 15 psi
2. First-class engineer	Age 21, 2 years operating on second-class license	100 to 150 boiler hp above 15 psi

TABLE 1-1 United States Stationary Engineer's License Requirements *(Continued)*

States and cities	Class license	Education, experience, and remarks	Plant capacity for class license
Mt. Vernon: Boiler Inspector, City Hall, Mt. Vernon, 10500	3. Chief engineer, unlimited	Age 21, 3 years operating on first-class license	Unlimited plant capacity
	1. Fireman	Age 21	Steam boiler or battery of boilers of 75 boiler hp (combined), or combined 825-sq-ft heating surface, 15 psi and evaporating 2,586 lb of water/hr from 212°F
	2. Second-class engineer	Age 21	Same as above, except boiler hp of up to 150 instead of 75
	3. First-class engineer, unlimited	Age 21	Unlimited plant capacity
New York City: Bureau of Examinations, Civil Service Comm., Municipal Bldg., New York, 10013	1. Fireman, with or without oil-burner endorsement	Age 21, 2 years as fireman, water tender, oiler, or assistant stationary or marine engineer	All boilers over 15 psi and 100-sq-ft heating surface require licensed operator, but class plant for license not specified
	2. Portable engineer	Age 21; 5 years as fireman, oiler, or assistant to licensed operating engineer in N.Y.C. within last 7 years; or 5 years as boilermaker or machinist, 1 year being under licensed engineer in N.Y.C. in last 3 years; or at least 1 year in stationary plant under N.Y.C. engineer while holding M.E. degree; or 1 year in sta-	

	tionary plant under N.Y.C. engineer while holding any engineer's license issued by U.S. Government, state, or territory	
3. Third-grade engineer (with or without oil-burner endorsement)	Same as above	
4. Second-grade engineer	Age 21, 2 years continuous work on third-grade N.Y.C. license	
5. First-grade engineer	Age 21, 1 year continuous work on second-grade N.Y.C. license	
Niagara Falls: Board of Examiners, Stationary Engineers, City Hall, Niagara Falls, 24302		
1. First-class fireman	Age 19, 6 months as helper in boiler room	All steam boilers
2. First-class engineer	Age 20, 1 year on fireman license	Not stipulated
3. Chief engineer	Age 21, 3 years on first-class engineer's license	Not stipulated
Rochester: City Public Safety, Bldg., Rochester, 14614		
1. Third-class engineer	Age 21, 1 year in steam plant	Any boiler or pressure vessel up to 100 boiler hp
2. Second-class engineer	Age 21, 2 years in steam stationary plant	Up to 500 boiler hp
3. First-class engineer	Age 21, 3 years in stationary steam plant	Up to 1500 boiler hp
4. Chief engineer	Age 21, first-class license and 5 years in power steam plants	Unlimited steam plant

TABLE 1-1 United States Stationary Engineer's License Requirements *(Continued)*

States and cities	Class license	Education, experience, and remarks	Plant capacity for class license
Tonawanda: Smoke Abatement Office, City Hall, Tonawanda, 14150	1. Special engineer	Age 21, must prove capability of operating a specific power plant	Limited to specific plant, up to 30 boiler hp and/or engine hp, up to 100 psi
	2. Second-class engineer	Age 21; 3 years firing boilers or operating engines, etc., or building power units; or technical school graduate	Up to 100 boiler hp
	3. First-class engineer	Age 21, 2 years on second-class license	Up to 225 boiler hp
	4. Chief engineer	Age 21, 3 years on first-class license	Power plant of any capacity
White Plains: Fire Dept., Municipal Bldg., White Plains, 10605	1. Second-class engineer	Age 21, 2 years under second-class engineer, or equivalent	High-pressure plant up to 75 boiler hp
	2. First-class engineer	Age 21, 3 years under first-class engineer, 1 year holding second-class license	High-pressure steam plant up to 200 boiler hp
	3. Chief engineer	Age 21, 4 years under chief engineer in high-pressure plant, and 1 year holding first-class license	Any high-pressure steam plant
Yonkers: Board of Examiners, City Hall, Yonkers, 10700	1. Fireman	Age 21	All boilers require licensed operators, but class plant for license not stipulated
	2. Portable engineer	Age 21	

			Up to 500 boiler hp
	3. Second-class engineer	Age 21	
	4. First-class engineer	Age 21	Unlimited
Ohio Examiners of Steam Engineers, State Capitol, Columbus, 43215	1. Low-pressure boiler operator	Age 21, 1 year as boiler room attendant, operating, maintaining, or erecting boilers	All boilers over 30 boiler hp but under 15 psi must have licensed operators; class plant for license not stipulated
	2. Boiler operator	Not specified	
	3. Third-class engineer	Age 21, 1 year as engineer, oiler, fireman, or water tender of steam boilers	
	4. Second-class engineer	Age 21, 2 years as licensed operating engineer	
	5. First-class engineer	Age 21, 2 years in steam plants as licensed operating engineer	
Oklahoma	No state license		
Oklahoma City: Boiler Inspector, City Hall, Oklahoma City, 73102	First-class engineer	Age 21, 2 years in steam plants under practical engineer	Unlimited capacity
Tulsa: Board of Examiners, 200 Civic Center, Tulsa, 74103	1. Fireman	1 year in steam plants	All heating boilers above 15 psi and 1,000,000 Btu, steam boilers above 15 psi and 100-sq-ft heating surface require licensed operator, but class plant on license not stipulated
	2. Third-class engineer	1 year in steam and refrigeration plants	

TABLE 1-1 United States Stationary Engineer's License Requirements (Continued)

States and cities	Class license	Education, experience, and remarks	Plant capacity for class license
	3. Second-class engineer	3 years in steam and refrigeration plants	Up to 15-psi boilers and up to 75-ton refrigeration and air-conditioning plants
	4. First-class engineer	5 years in steam and refrigeration plants	Unlimited
Panama Canal Zone Cristobal, Canal Zone	Yes, but not stipulated		
Pennsylvania	No state license		
Erie: Bureau of Licenses, Municipal Bldg., Erie, 16501	1. Class 3 water tender	Age 21, 1 year with steam-generating equipment, high school grad; tech school degree in lieu 1 year experience	Up to 100-hp steam-driven machinery, 2,000-sq-ft heating surface boiler; any size power equipment under licensed engineer
	2. Class 2 stationary engineer	Age 21, 2 years holding water tender license	Up to 200-hp steam-driven machinery, 4,000-sq-ft heating-surface boiler; any size power equipment under licensed chief engineer
	3. Class 1 chief stationary engineer	Age 21, 2 years holding class 2 license	Unlimited
Philadelphia: Bureau of Licenses & Inspections, Municipal Services Bldg., Philadelphia, 19107	1. Grade D fireman	Age 21, 2 years in boiler room	Over 15 psi, 30-boiler-hp steam prime movers; 25-ton refrigeration equipment
	2. Grade C portable and stationary engineer	Age 21, 2 years assistant engineer or helper	Not specified

License class	Requirements	Scope	Remarks
3. Grade A stationary engineer	Age 21, 2 years assistant engineer or helper	Unlimited	
Pittsburgh: Bureau Bldg. & Inspections, 100 Grant St., Pittsburgh, 15219			
1. Fireman	Age 21, 2 years assistant engineer or helper		All boilers and pressure vessels require licensed operators, but license class for plant not specified
2. Portable engineer	Age 21, 2 years as engineer, oiler, fireman of steam boilers	Not specified	
3. Stationary engineer	Not specified	Unlimited	
Rhode Island	No state license		
Providence: Dept. Public Service, City Hall, Providence, 02903			
1. Boiler operator, fireman, and water tender	Age 18, 6 months apprentice under licensed operator while holding permit		All boilers over 30 boiler hp, refrigeration over 15 tons require licensed operator, but class license for plant not specified
2. Stationary engineer, unlimited	Age 21; 5 years boiler operator; or 3 years assistant to licensed engineer; or 1 year in steam plants while holding permit if degreed engineer		
Woonsocket: Licensing Steam Engineer, City Hall, Woonsocket, 02895			
1. Boiler operator, fireman, and water tender	Age 21, 6 months apprentice under licensed operator while holding permit		All boilers over 30 boiler hp and refrigeration over 15 tons require licensed operator, but class license for plant not specified
2. Stationary engineer	Age 21; 5 years boiler operator; or 3 years assistant to licensed operator; or 1 year in steam plants while holding permit	Unlimited	

States and cities	Class license	Education, experience, and remarks	Plant capacity for class license
South Carolina	No state license		
South Dakota	No state license		
Tennessee	No state license		
Memphis: Safety Engineer Division, 125 N. Main St., Memphis, 38103	1. Third-grade engineer	Age 21, 3 years in steam plants	Up to 50 boiler hp, 50-hp compressor, 40-ton refrigeration, or assist second- or first-grade engineer
	2. Second-grade engineer	Same as above	Up to 100 boiler hp, 100-hp internal combustion engine, 100-hp compressor, 100-ton refrigeration, or assist first-grade engineer
	3. First-grade engineer	Age 21, hold third- and second-grade licenses	Unlimited
Texas	No state license		
Houston: Public Works Dept. Bldg., Houston, 77001	1. Boiler operator	Age 20, 2 years water tender, oiler, fireman, or 6 months as above if degreed engineer	Low-pressure hot-water-heating boilers, high-pressure steam boilers up to 1,004,000 Btu
	2. Third-grade engineer	Age 20, 2 years licensed engineer, oiler, water tender, boiler repairman, fireman, or 6 months of above if degreed engineer	Up to 1,674,000 Btu boiler, or act as shift engineer of up to 6,696,000 Btu under second-grade engineer
	3. Second-grade engineer	Age 21, 3 years stationary engineer, oiler, water tender, fireman, if 1 of 3 years under licensed operator, or if degreed engineer	Up to 6,696,000 Btu boiler, or unlimited capacity under licensed operator

Jurisdiction	License	Requirements	Scope
	4. First-grade engineer	Age 23; 5 years stationary engineer, oiler, water tender, fireman if 2 of 5 years as licensed engineer; or graduate engineer with 2 years in above	Unlimited
Utah	No state license		
Salt Lake City: Power and Heating Div., City & County Bldg., Salt Lake City, 84111	1. Second-class fireman	Age 18, 1 year in steam plants	Low-pressure boilers up to 50 boiler hp
	2. First-class fireman	Age 18, 2 years in steam plants	Low- or high-pressure boilers, unlimited capacity
	3. Second-class engineer	Age 20, 3 years in steam plants	Low- or high-pressure plant up to 100 hp
	4. First-class engineer	Age 20, 4 years in steam plants	Any capacity above 100 hp
Vermont	No state license		
Virginia	No state license		
Washington	No state license		
Seattle: Civil Service Dept., City Hall, Seattle, 98108	1. Grade 5 fireman	Age 21, no experience necessary	Up to 15 psi and up to 500-sq-ft heating surface
	2. Grade 4 fireman	Age 21, 3 years in steam power plant	Boilers of any capacity
	3. Grade 3 engineer	Age 21; 3 years in steam plant; or 1 year in steam plant after serving 3 years in steam engine works; or 1 year in steam plants if degree from school of technology	Up to 250 boiler hp
	4. Grade 2 engineer	Same as above	Up to 1,500 boiler hp

TABLE 1-1 United States Stationary Engineer's License Requirements (Continued)

States and cities	Class license	Education, experience, and remarks	Plant capacity for class license
	5. Grade 1 engineer	Same as above	Unlimited
Spokane: Dept. of Buildings, City Hall, Spokane, 99201	1. Fireman	Age 18	Steam boiler up to 300-sq-ft heating surface or heating boiler up to 15 psi
	2. Third-class engineer	Age 21, 1 year as steam engineer	Up to 100 boiler hp, or shift engineer under second-class engineer
	3. Second-class engineer	Age 21, 1 year as steam engineer	Up to 200 boiler hp, or shift engineer under first-class engineer
	4. First-class engineer	Age 21, 2 years on steam license, or 1 year steam engineer while holding M.E. degree	Unlimited
Tacoma: Boiler Inspector, County-City Bldg., Tacoma, 98402	1. Low-pressure fireman	Age 18, produce satisfactory evidence of qualification	Up to 15 psi steam boiler or hot-water boiler up to 160 psi and 250°F
	2. High-pressure fireman	Age 21, 2 years as licensed low-pressure fireman, or technology school graduate	Up to 5,000-sq-ft heating surface of unlimited pressure if used as auxiliary, or plant of unlimited capacity under licensed engineer
	3. Engineer	Age 21; 2 years high-pressure fireman; or 3 years licensed low-pressure technology school graduate	Unlimited steam plant
	4. Chief engineer	Age 23, 5 years with high-pressure boilers, or technology school graduate holding engineer license for 2 years	Unlimited steam plant
West Virginia	No state license		

Wisconsin — No state license

	License	Requirements	Scope
Kenosha: Board of Examiners, City Hall, Kenosha, 53140	1. Third-class engineer	Age 21, 2 years in steam plants	Up to 50 boiler hp
	2. Second-class engineer	Age 21, 2 years in steam plants	Up to 150 boiler hp
	3. First-class engineer	Age 21, 2 years in steam plants	Unlimited
Milwaukee: Dept. Bldg. Inspection & Safety Engineering, Municipal Bldg., Milwaukee, 53202	1. Low-pressure fireman	Age 21, 1 year in steam boiler plants	Operate up to 3 low-pressure plants (within 3-block area) of up to 10 boiler hp
	2. High-pressure fireman	Age 21, 1 year in steam boiler plants	Up to 75 boiler hp but no steam engines except boiler auxiliaries
	3. Third-class engineer	Age 21, 2 year in high-pressure steam plants	Up to 75 boiler hp and steam engines
	4. Second-class engineer	Age 21, 3 years in high-pressure steam plants	Up to 300 boiler hp
	5. First-class engineer	Age 21, 4 years in high-pressure steam plants	Unlimited
Racine: Stationary Engineer Examiner, City Hall, Racine, 53203	1. Third-class engineer	Age 19, course of instruction in steam boiler operation	Up to 75 boiler hp, or act as assistant to second-class engineer
	2. Second-class engineer	Age 21, 2 years in steam boiler plants	Up to 300 boiler hp, or act as assistant to first-class engineer
	3. First-class engineer	Age 21, 2 years in steam boiler plants	Unlimited steam-generating plant

Wyoming — No state license

TABLE 1-2 Canadian Stationary Engineer's License Requirements

Province	Class license	Education, experience, and remarks	Plant capacity for class license
Alberta Dept. Manpower & Labor, Boiler Branch, IBM Bldg., Edmonton, T5K OG5	1. Fireman	Age 18, 6 months as boiler operator	Up to 50 boiler hp, or in charge of shift of 100 boiler hp, or watchman of steam plant
	2. Fourth-class engineer	Age 19; 1 year as fireman or engineer in high-pressure plant of up to 10 boiler hp; or 6 months practical experience if holding M.E. degree; or 2 years as fireman with certificate from technical school	Chief of steam plant up to 100 boiler hp, or shift engineer up to 500 boiler hp, or assistant engineer up to 1,000 boiler hp
	3. Third-class engineer	Age 20; 1 year as chief steam engineer in high-pressure plant over 50 boiler hp while holding fourth-class license; or 1 year as shift engineer in high-pressure plant above 100 boiler hp while holding fourth-class license; or 1 year as assistant engineer in high-pressure plant above 500 boiler hp before holding fourth-class license; or 30 months as fireman in high-pressure plant above 100 boiler hp under licensed engineer; or have M.E. degree and one-half of above time operating; or 1 year building or erecting machinery and one-half above operating experience	Chief of steam plant up to 500 boiler hp, or shift engineer of up to 1,000 boiler hp, or assistant engineer of any capacity
	4. Second-class engineer	Age 21; hold third-class license and 2 years as chief of high-pressure plant above 100 boiler hp; or 2 years	Chief of steam plant up to 1,000 boiler hp, or shift engineer of any capacity

	as shift engineer in high-pressure plant above 500 boiler hp; or 3 years as shift engineer of high-pressure plant above 100 boiler hp; or 2 years assistant engineer in high-pressure steam plant above 1,000 boiler hp; or 2 years of above and hold M.E. degree.	Unlimited
5. First-class engineer	Age 21; 3 years as chief in high-pressure plant above 500 boiler hp; or 3 years as shift engineer in high-pressure plant above 1,000 boiler hp; or 4 years as shift engineer in high-pressure plant above 500 boiler hp; or 4 years as assistant shift engineer in high-pressure plant above 5,000 boiler hp; or four other qualifications too detailed to publish here	
6. Special oil-well certificate	6 months with oil-well equipment	Up to 100 boiler hp boilers on drilling site
7. Temporary certificate	Not stipulated	Chief steam engineer, assistant engineer, or fireman of plant indicated on certificate, or operate boilers on drilling site
British Columbia Dept. Public Works, 501 W. 12th Ave., Vancouver, V5Z 1M4	1. Boiler operator	6 months fireman, utility or engineer, or 24 months apprentice or journeyman in allied trades; vocational training required	Up to 50 boiler hp high-pressure, up to 200 boiler hp low-pressure; or shift engineer up to 100 boiler hp high-pressure, up to 300 boiler hp low-pressure

TABLE 1-2 Canadian Stationary Engineer's License Requirements (Continued)

Province	Class license	Education, experience, and remarks	Plant capacity for class license
	2. Fourth-class engineer	1 year fireman, utility or engineer, or 4 years apprentice or journeyman in allied trades; vocational training required	Up to 100 boiler hp high-pressure, up to 300 boiler hp low-pressure; or shift engineer up to 450 boiler hp high-pressure, and any low-pressure plant
	3. Third-class engineer	1 year in charge of high-pressure plant above 50 boiler hp; or shift engineer in high-pressure plant over 500 boiler hp; or professional engineer with 6 months in charge; or shift engineer of above; or 4 years as process operator or journeyman in allied trades	Up to 450 boiler hp high-pressure; or any low-pressure plant, or shift engineer any plant up to 1,000 boiler hp
	4. Second-class engineer	Hold third-class certificate; or 2 years chief engineer of high-pressure plant above 100 boiler hp; or shift engineer above 500 boiler hp; or 3 years as shift engineer in high-pressure plant over 100 boiler hp (periods above cut in half if 2 years in allied trade or journeyman), or professional engineer with 1 year as shift engineer above 500 boiler hp	Shift engineer in any plant

5. First-class engineer	Hold second-class engineer certificate and 30 months as chief engineer of high-pressure plant above 500 boiler hp; or 30 months shift engineer above 1,000 boiler hp; or 42 months as assistant shift engineer above 1,000 boiler hp (periods above cut to 18 months for chief engineer duty if journeyman for 36 months); or professional engineer holding second-class with 18 months as shift engineer in plant above 1,000 boiler hp. NOTE: Recognition is given to marine certificates; credits allowed for experience in related fields if inspector feels applicant will do well in steam plants.	Unlimited
Manitoba Mechanical & Engineering Div., 611 Norquay Bldg., Winnipeg. R3C OP8		
1. Fireman	Age 18, 3 months in low-pressure heating plant	50- to 200 boiler hp low-pressure steam heating plant
2. Fourth-class engineer	Age 18, 1 year with high-pressure boiler over 100 boiler hp, or 2 years with low-pressure boiler above 200 boiler hp	Above 200 boiler hp low-pressure heating boilers
3. Third-class engineer	Age 18; hold fourth-class certificate and 18 months chief in 100- to 300-boiler-hp plant; or 2 years shift engineer in 300 to 750-boiler-hp plant; or 3 years in plant above 300 boiler hp	100- to 300-boiler-hp high-pressure steam boiler plant

Province	Class license	Education, experience, and remarks	Plant capacity for class license
	4. Second-class engineer	Age 18; hold third-class certificate and 18 months chief in 100- to 300-boiler-hp high-pressure plant; or 2 years shift engineer in 300- to 750-boiler-hp high-pressure plant; or 3 years in plant of over 300 boiler hp, high-pressure	300- to 750-boiler-hp high-pressure steam boiler plant
	5. First-class engineer	Age 18; hold second-class certificate and 2 years chief in 300 to 750-boiler-hp high-pressure plant; or 30 months shift engineer in plant above 750-boiler-hp high-pressure; or 3 years in plant above 750 boiler hp	Unlimited
New Brunswick Dept. of Labor, P.O. Box 580, Fredericton, E3B 5H1	1. Fourth-class engineer	Age 18, 1 year in heating plant or power plant	Any low-pressure heating plant; or high-pressure heating plant up to 300 boiler hp; or power plant up to 300 boiler hp; or have charge of low-pressure heating plant up to 200 boiler hp; or have charge of high-pressure heating plant up to 100 boiler hp
	2. Third-class engineer	Age 18, 2 years in heating or power plant, or 1 year while holding fourth-class license	Any low-pressure heating plant; or power plant up to 750 boiler hp; or charge of any low-pressure heating plant; or charge of high-pressure heating plant up to 300 boiler hp; or charge of power plant up to 300 boiler hp

	Requirements	Scope
3. Second-class engineer	Age 18, hold third-class license 2 years, and total 4 years in heating or power plant, 1 year being in heating or power plant above 300 boiler hp	Any low-pressure heating plant and any high-pressure heating or power plant, or have charge of any low-pressure or high-pressure heating or power plant up to 750 boiler hp
4. First-class engineer	Age 18, hold second-class license 2 years with 6 years in heating or power plant, 2 years being in heating or power plant above 750 boiler hp	Unlimited steam power plant
Newfoundland and Labrador Engineering & Technical Services Div., Confederation Bldg., St. John's, A1C 5T7		
1. Fireman	Age 19, 1 year high-pressure boiler above 15 boiler hp, or low-pressure boiler above 50 boiler hp on boiler room shift	Stationary low-pressure heating plant up to 100 boiler hp; or shift fireman of high-pressure heating plant up to 200 boiler hp. or low-pressure heating plant up to 400 boiler hp
2. Fourth class, grade AB	Age 19, 1 year high-pressure heating plant and steam engine or steam turbine plant above 25 boiler hp each	(a) Chief engineer of: stationary combined-pressure plant up to 200 boiler hp, or high-pressure heating plant up to 200 boiler hp and 200 engine hp, or low-pressure heating plant up to 400 boiler hp, or refrigeration plant up to 400 hp, or compressed gas plant up to 400 hp, or portable plant of unlimited hp. (b) Shift engineer of: combined-pressure plant up to 400 hp, or high-pressure heating plant up to 400 boiler hp, or low-pressure heating plant up to 800 hp, or refrigeration plant unlimited, or compressed gas plant of unlimited hp, or portable plant unlimited.

TABLE 1-2 Canadian Stationary Engineer's License Requirements *(Continued)*

Province	Class license	Education, experience, and remarks	Plant capacity for class license
	3. Fourth-class, grade A	Age 19, 1 year in combined-pressure plant, high-pressure heating plant, or on shift in high-pressure boiler room above 25 boiler hp, or low-pressure boilers above 75 boiler hp	(a) Chief engineer of: stationary combined-pressure plant up to 200 hp, or high-pressure heating plant up to 200 boiler hp, or low-pressure heating plant up to 400 boiler hp, or refrigeration up to 400 hp, or compressed-gas plant unlimited, or portable plant unlimited. (b) Shift engineer of: combined-pressure plant up to 400 hp, or high-pressure heating plant up to 400 boiler hp, or low-pressure heating plant up to 800 boiler hp, or refrigeration unlimited, or compressed gas unlimited, or portable plant unlimited.
	4. Third-class, grade AB	Qualification for fourth-class, Grade AB plus 1 year in high-pressure heating plant above 100 boiler hp and 100-hp steam engine-turbine, 9 months being on regular shift in high-pressure plant of 100 hp and hold fourth-class, grade AB license	(a) Chief engineer of: stationary combined-pressure plant up to 400 boiler hp, or high-pressure heating or power plant up to 400 boiler hp and 400 engine hp, or low-pressure heating plant of 800 boiler hp. (b) Shift engineer of: stationary combined-pressure plant up to 800 hp, or high-pressure heating or power plant up to 800 hp, or low-pressure heating plant unlimited hp.

Class	Qualification	Scope
5. Third-class, grade A	Qualification for fourth-class, grade A plus 1 year in stationary combined-pressure plant or high-pressure heating plant above 100 boiler hp, 9 months being on regular shift in boiler room on high-pressure boilers above 100 boiler hp and hold fourth-class, grade A license	(a) Chief engineer of: stationary combined-pressure plant up to 400 hp, or high-pressure heating plant up to 400 boiler hp, or low-pressure heating plant up to 800 boiler hp, or stationary refrigeration unlimited, or compressed-gas plant unlimited. (b) Shift engineer of combined-pressure up to 800 hp, or high-pressure heating plant up to 800 boiler hp, or low-pressure heating plant unlimited hp.
6. Second-class, grade AB	Qualification as third-class, grade AB plus 2 years in high-pressure heating plant and power plant above 200 boiler hp, and 200 engine-turbine hp, 1 year being on regular shift in boiler-engine room of high-pressure boilers and steam engines or turbines each above 200 hp and hold third-class, grade AB licenses	(a) Chief engineer of: combined-pressure plant up to 800 hp, or high-pressure heating plant up to 800 boiler hp and 800 engine hp, or low-pressure heating plant. (b) Shift engineer of: combined-pressure plant unlimited hp, or high-pressure heating or power plant unlimited hp, or low-pressure heating plant unlimited hp.
7. Second-class, grade A	Qualification for third-class, grade A and 2 years combined-pressure plant or high-pressure heating plant up to 200 boiler hp, 1 year being on shift in boiler room on high-pressure boilers above 200 boiler hp and hold third-class, grade A license	(a) Chief engineer of: combined-pressure plant up to 800 hp, or high-pressure heating plant up to 800 boiler hp, or low-pressure heating plant unlimited. (b) Shift engineer of: combined-pressure plant unlimited hp, or high-pressure heating plant unlimited hp, or low-pressure heating plant unlimited hp.

TABLE 1-2 Canadian Stationary Engineer's License Requirements (Continued)

Province	Class license	Education, experience, and remarks	Plant capacity for class license
	8. First-class, grade AB	Qualification as second class, plus 2 years in high-pressure heating and power plant above 400 boiler hp and 400 steam engine-turbine hp, 1 year being on regular shift in boiler-engine room, on high-pressure boilers, engines, or turbines, each above 400 hp, or 2 years as inspector and hold second-class, grade AB license	Chief engineer or shift engineer of high-pressure heating-plant of un-limited hp
	9. First-class, grade A	Qualification as second-class plus 2 years combined-pressure plant; or high-pressure heating plant above 400 boiler hp, 1 year being on shift in boiler room high-pressure boilers total hp above 400 boiler hp; or 2 years as inspector and hold second-class, grade A license	Chief engineer or shift engineer of combined-pressure plant, or high-pressure heating plant, or low-pressure heating plant unlimited boiler hp
Northwest Territories Boilers and Pressure Vessels Inspection Branch, Dep't of Public Services Yellowknife, XOE 1HO	1. Class 5 engineer	All examinations are under the Canadian Standard Plan, with same conditions as indicated for British Columbia, Alberta, Manitoba, and Saskatchewan	Take charge of low-pressure boiler or any pressure vessel or plant up to 50 hp, or operate up to 200 hp under class 3 engineer
	2. Class 4 engineer		Take charge of any boiler or pressure vessel up to 200 hp, or operate up to 500 hp under Class 3 engineer

44

Classification	Requirements	Duties
3. Class 3 engineer		Take charge of any boiler, pressure vessel, or plant up to 500 hp as chief engineer, or operate any plant up to 750 hp as shift engineer under chief engineer
4. Class 2 engineer		Take charge of any plant up to 750 hp as chief engineer, or operate any plant as shift engineer under chief engineer
5. Class 1 engineer		Unlimited steam plant
Nova Scotia Board of Examiners, 5668 South St., P.O. Box 697, Halifax, B3J 2T8 1. Engine operator, fourth class	Age 18, grade 8 graduate, 9 months in boiler plant of over 25 boiler hp	Low-pressure boiler up to 200 boiler hp; high-pressure boiler up to 100 boiler hp, combined boiler plant up to 100 boiler hp; steam engine plant, steam turbine plant, or compressor plant up to 100 hp
2. Engine operator, third class	Age 19, grade 8 graduate, 18 months in high-pressure plant boiler above 75 boiler hp	Low-pressure boiler up to 600 boiler hp, high-pressure boiler up to 200 boiler hp, combined boiler plant up to 200 boiler hp, steam engine plant up to 200 boiler hp, steam turbine plant up to 200 hp, compressor plant up to 400 hp
3. Engine operator, second class	Grade 10 graduate, 54 months in high-pressure steam plant including (a) 36 months in high-pressure boiler plant over 200 boiler hp, (b) 18 months in steam engine and/or steam turbine plant over 200 machine hp	Low-pressure boiler unlimited boiler hp, high-pressure boiler up to 600 boiler hp, combined boiler plant up to 600 boiler hp, steam engine plant up to 600 boiler hp, steam turbine plant up to 600 hp, compressor plant unlimited hp

TABLE 1-2 Canadian Stationary Engineer's License Requirements (Continued)

Province	Class license	Education, experience, and remarks	Plant capacity for class license
	4. Engine operator, first class	Grade 11 graduate, 72 months in high-pressure steam plant, including (a) 36 months in low-pressure boiler plant over 400 boiler hp, (b) 18 months in high-pressure boiler plant over 600 hp, (c) 18 months in steam engine and/or steam turbine plant over 500 hp	Unlimited capacity for following plants: low-pressure boiler plant, high-pressure boiler plant, combined boiler plant, steam engine plant, steam turbine plant, compressor plant
Ontario Industrial Training Branch, 400 University Ave., Toronto, M5G 1S5	1. Fourth-class engineer	Age 18; 1 year in power plant over 17 therm-hours (therm-hr = 100,000 Btu/hr, hp = 33,470 Btu/hr) or 50 boiler hp; or low-pressure plant of 50 therm-hr (150 hp). Only 3 months qualifying experience required in accreditation under modular system.	Take charge high-pressure plant up to 150 hp, or low-pressure plant up to 400 hp.
	2. Third-class engineer	2 years over 17 therm-hr; or low-pressure plant over 50 therm-hr plus hold fourth-class license	Take charge high-pressure plant up to 134 therm-hr or low-pressure plant up to 400 therm-hr
	3. Second-class engineer	3½ years total, of which 2 years for third-class and 1½ years over 134 therm-hr plus hold third-class license	Take charge of high-pressure plant up to 400 therm-hr, or any low-pressure plant
	4. First-class engineer	6 years total, of which 2 years for second-class above 300 therm-hr, plus hold second-class license NOTE: see Appendix for more information on therm-hr	Unlimited

Prince Edward Island Dept. of Labor, Box 2000, Charlotte-town, C1A 7N8	1. Fireman	Age 18, 3 months in heating plants	Fireman in high-pressure heating plant up to 50 boiler hp, or fireman in low-pressure heating plant up to 100 boiler hp
	2. Fourth-class power engineer	Age 18, 1 year in heating plant or power plant. NOTE: Special engineering training or course in power engineering, or repair or construction of boilers, or license from some other province, state, or country will be considered in lieu of practical experience.	Chief engineer in low-pressure heating plant up to 200 boiler hp, or chief engineer in high-pressure heating plant up to 100 boiler hp, or shift engineer in any low-pressure plant, or shift engineer in high-pressure heating or power plant up to 300 boiler hp, or assistant engineer in high-pressure heating plant or power plant up to 750 hp
	3. Third-class power engineer	Age 18, 2 years in heating or power plant, or 1 year while holding fourth-class license	Chief engineer in any low-pressure heating plant, or chief engineer in high-pressure heating plant up to 300 boiler hp, or shift engineer in any low-pressure heating plant, or shift engineer in high-pressure heating or power plant up to 750 hp, or assistant engineer in any heating or power plant
	4. Second-class power engineer	Age 18, hold third-class license for 2 years, plus 4 years in heating or power plants, 1 year of which in plant over 300 hp	Chief engineer in any low-pressure heating plant, or chief engineer in high-pressure heating plant up to 750 boiler hp, or shift engineer in any plant

TABLE 1-2 Canadian Stationary Engineer's License Requirements *(Continued)*

Province	Class license	Education, experience, and remarks	Plant capacity for class license
	5. First-class power engineer	Age 18, hold second-class license for 2 years plus 6 years in heating and power plants, of which 2 years in plants above 750 hp	Unlimited capacity
Quebec Board of Examiners, Stationary Enginemen, 255 Cremazie East, Montreal, G1R 4Z1	1. Fifth-class, heating-steam engines	4 months in heating or steam engine plant, or fabrication, installation, or repair of steam engines and boilers	Any heating or steam plant up to 100 hp, or fourth- or lower-class plant for one day, or any electric boilers up to 200 boiler hp
	2. Fourth-class, heating-steam engines	Hold fifth-class license and serve with such for: 2 years in fifth-class plant, or 9 months in fourth-class or higher plant, or 6 months in fifth-class plant; or hold fourth- or higher-class license in marine, heating, or steam engines for past 1 year; or complete technical courses; or work under fourth-class or higher engineer for: (a) 1 year as helper in fourth-class plant, (b) 10 months as helper in fourth-class plant plus completed technical courses, (c) 8 months as helper in fourth-class plant and completed technical courses	Any heating or steam plant up to 300 hp, or third-class plant for one day, or electric boiler plant up to 400 boiler hp
	3. Third-class, heating-steam engines	Hold fourth-class license and serve under third-class operator for one of the following periods: (a) 1 year third-class plant, (b) 10 months in	Any heating or steam plant up to 600 hp, or second-class plant for one day, or electric boiler plant up to 600 boiler hp

	third-class plant plus completing technical courses, (c) 8 months in third-class plant plus completed technical courses; or hold third-class license in marine heating or steam engines for past 1 year; or complete 3-year course in technical school	
4. Second-class, heating-steam engines	Hold third-class license and serve under second-class operator in heating and steam engine plant for one of the following periods: (a) 18 months in second-class plant, (b) 16 months in second-class plant, plus completed technical courses, (c) 14 months in second-class plant plus completed technical courses; or hold second-class license for marine, heating, or steam engines for 1 year	Any heating or steam engine plant up to 1,000 hp, or any electric boiler plant
5. First-class, heating-steam engines	Hold second-class license in heating and steam engines plus serving under first-class operator for: (a) 2 years in first-class plant of which 1 year in charge, (b) 22 months in first-class plant of which 1 year in charge plus completed technical courses, (c) 20 months in first-class plant of which 1 year in charge plus completed technical courses; or hold first-class license in heating and marine steam engines for 1 year	Unlimited capacity

TABLE 1-2 Canadian Stationary Engineer's License Requirements *(Continued)*

Province	Class license	Education, experience, and remarks	Plant capacity for class license
Saskatchewan Boiler & Pressure Vessel unit 1150 Rose St., Regina, S4P 2YR	1. Fireman	Age 16, satisfy inspector on knowledge and experience to operate and repair boilers	Low-pressure boiler or steam plant up to 100 hp and/or high-pressure boiler for heating only and up to 30 boiler hp, or assist holder of fifth-class license
	2. Special certificate	Age 18, satisfy inspector on knowledge and experience to operate and repair boilers for plant for which certificate is issued	Steam plant indicated in certificate up to 50 boiler hp
	3. Fifth-class engineer	Age 17; hold fireman license 1 year in plant with over 30 hp boiler; or high-pressure plant over 5 hp; or assist in plant over 100 hp	Low-pressure boiler or heating plant, or high-pressure boiler up to 50 boiler hp, or assist in plant up to 150 hp
	4. Fourth-class engineer	Any person; (a) 1 year operate high-pressure boiler up to 25 boiler hp, or (b) 1 year assist with high-pressure boiler above 100 boiler hp, or (c) hold fifth-class license plus 2 years as chief in low-pressure plant above 75 boiler hp, or (d) hold fifth-class license plus 2 years assisting in low-pressure plant above 150 hp, or (e) complete power engineering course, or (f) graduate M.E., or (g) one half time in (a) and (b) plus 1 year with equipment indicated. 6 months credit for completed course in power engineering.	Chief engineer of boiler or steam plant up to 100 hp, or assist in boiler or steam plant up to 500 hp

5. Third-class engineer	Fourth-class license and (*a*) 1 year as chief engineer in high-pressure plant above 50 hp, or (*b*) 1 year shift engineer in high-pressure steam plant above 100 hp, or (*c*) 1 year assistant shift engineer in high-pressure steam plant over 500 hp, or (*d*) 2 years chief engineer or shift engineer in low-pressure steam plant above 300 hp, or (*e*) one-half time in (*a*), (*b*), (*c*), or (*d*) plus 1 year of design, construction, repair, etc., of equipment indicated. Or degreed engineer with 6 months in high-pressure steam plants above 100 hp, or 6 months credit for completion of approved course.	Chief engineer in boiler or steam plant up to 500 hp, or assist in plant up to 1,000 hp
6. Second-class engineer	Third-class license plus (*a*) 2 years chief of high-pressure plant above 100 hp, or (*b*) 2 years shift engineer of high-pressure plant above 500 high-pressure, or (*c*) 3 years shift engineer in high-pressure above 100 hp, or (*d*) 2 years assisting shift engineer in high-pressure plant above 1,000 hp, or (*e*) 1 year of (*a*), (*b*), (*c*), or (*d*) plus engineer degree, or (*f*) one-half time in (*a*), (*b*), (*c*), or (*d*) plus 2 years of design, construction, repair, maintenance, etc., of power plant equipment. 9 months credit for completed power engineer course.	Chief engineer in boiler or steam plant up to 1,000 hp, or assist in boiler or steam plant of any capacity

TABLE 1-2 Canadian Stationary Engineer's License Requirements *(Continued)*

Province	Class license	Education, experience, and remarks	Plant capacity for class license
	7. First-class engineer	Hold second-class license plus (a) 30 months as chief engineer of high-pressure plant above 500 hp, or (b) 30 months as shift engineer in high-pressure plant above 1,000 hp, or (c) 42 months assisting shift engineer in high-pressure plant above 1,000 hp, or (d) 15 months of (a), (b), or (c) if graduate engineer, or (e) one-half time in (a), (b), or (c) plus 3 years of design, construction, repair, etc., of power equipment. 1 year credit for completed power engineering course.	Unlimited boiler or steam plant
Yukon Territory Box 2703, Whitehorse Y1A 2C6		Government of the Yukon Territory has an arrangement with the Province of British Columbia whereby the British Columbia standards are used. Examinations are requested, written in the Yukon, and forwarded to British Columbia for marking. The subsequent operator's license is mailed to those who qualify. See British Columbia for license requirements.	

TABLE 1-3 U.S. Merchant Marine Engineer's License Requirements

Apply to	Class license	Examination required	Education and experience	Citizenship	Nonlocal licenses recognized	Renewal	Remarks
Marine Inspection, U.S. Coast Guard, any U.S. port	Third assistant engineer; steam vessels, unlimited	Yes, written	1. Age 19, at least 3 years in engine department, 2 years and 6 months of which must be as fireman, oiler, water tender, or other qualified member of engine department: one-third of above service may have been on diesel ships; or 2. 3 years as machinist apprentice in construction or repair of marine, locomotive, or stationary engines and 1 year in engine department of steam vessels as oiler, water tender, or junior engineer; 1 year of this service may have been on diesel vessels; or 3. Graduate form: U.S. Merchant Marine Academy (engineering), State Nautical Schoolship (engineering), U.S. Naval Academy, or U.S. Coast Guard Academy; or 4. Completion of U.S. Maritime Service engineering course accepted as 4 months time; or	U.S. citizen	1. Experience on foreign vessels given due credit 2. (Stationary engineers take note.) When an applicant presents evidence of service or experience which does not meet the specific requirements of the regulations in this part, but which, in the opinion of the Officer in Charge of Marine Inspection, is a reasonable equivalent, the application for license with supporting data shall be submitted to the Commandant for evaluation, together with the recommendations of the Officer in Charge of Marine Inspection	No charge, renewal, every 5 years	1. No candidate for original license shall be examined until he presents a certificate from the U.S. Public Health Service that he has passed satisfactorily examination based on the contents of "The Ship's Medicine Chest and First Aid at Sea." 2. All applicants for original license must pass physical examination given by medical officer of U.S. Health Service; this certificate shall attest to applicant's acuity of vision, color sense, and general physical condition. 3. Applicants for original licenses shall be examined only for ability to distinguish colors red, blue, green, and yellow. 4. For original license, applicant must have, either with or

Apply to	Class license	Examination required	Education and experience	Citizenship	Nonlocal licenses recognized	Renewal	Remarks
			5. Graduate from marine engineering school of technology with 3 months in engine department of steam vessel; 1 month of this service may be on diesel vessels; Or 6. Graduate in mechanical or electrical engineering of recognized school with 6 months service in engine department of steam vessels, 2 months of which may be on diesel ships; Or 7. 1 year service as oiler, water tender, or junior engineer on steam vessels while holding license as third assistant engineer of diesel vessels				without glasses, at least 20/30 vision in one eye and at least 20/50 in the other; applicant who wears glasses must pass test without glasses of at least 20/50 in one eye and at least 20/70 in other. 5. No horsepower limitation shall be placed on a license where all of applicant's qualifying experience has been on ocean-going merchant vessels of 2,500 hp and over.
	Second assistant engineer; steam vessels, unlimited	Yes, written	1. Age 21, at least 1 year in charge of a watch, while holding license as third assistant engineer of steam vessels; or 2. 2 years as assistant to the engineer in charge of a watch, while holding license as third assistant engineer of steam vessels; or	U.S. citizen	Experience on foreign vessels given due credit	Every 5 years, no charge	

3. While holding license as second assistant engineer of diesel vessels, 6 months as third assistant engineer of steam vessels, 6 months as observer second assistant engineer on steam vessels; or 1 year as oiler, water tender, or junior engineer of steam vessels

First assistant engineer; steam vessels, unlimited

Yes, written

1. Age 21, 1 year as second assistant engineer of steam vessels; or
2. 2 years as third assistant or junior second assistant engineer in charge of a watch on steam vessels, while holding license as second assistant engineer of steam vessels; or
3. While holding license as first assistant engineer of diesel vessels: 6 months as second assistant engineer of steam vessels, 6 months as observer first assistant engineer of steam vessels, or 1 year as oiler, water tender, or junior engineer of steam vessels; or
4. 3 years as oiler, water tender, or fireman on steam vessels while holding license as first assistant of steam vessels of not more than 1,000 hp

U.S. citizen

Experience on foreign vessels given due credit

Every 5 years, no charge

TABLE 1-3 U.S. Merchant Marine Engineer's License Requirements *(Continued)*

Apply to	Class license	Examination required	Education and experience	Citizenship	Nonlocal licenses recognized	Renewal	Remarks
	Chief engineer; steam vessels, unlimited	Yes, written	1. Age 21, 1 year as first assistant engineer of steam vessels; or 2. 2 years as second assistant steam or junior first assistant engineer in charge of a watch on steam vessels while holding license as first assistant engineer of steam vessels; or 3. While holding license as chief engineer of diesel vessels: 6 months as first assistant engineer of steam vessels, 6 months as observer chief engineer of steam vessels, or 1 year as oiler, water tender, or junior engineer of steam vessels	U.S. citizen	Experience on foreign vessels given due credit	Every 5 years, no charge	
	Qualified member of the engine department (QMED), refrigerating engineer, oiler,	Yes, oral or written	1. Age 17, with consent of parents if not conflicting with labor laws of state	Does not have to be U.S. citizen		Good for life or until revoked	1. QMED is any person below rating of licensed officer and above rating of coal passer or wiper who holds certificate of service as QMED issued by Coast Guard or predecessor or authority.

water tender, deck engineer, junior engineer, electrician, boiler maker, machinist, pumpman

2. 6 months service at sea in rating at least equal to coal passer or wiper in engine department of vessel required to have such certified men, or in engine department of tugs or towboats operating on high seas or Great Lakes, or on bays or sounds directly connected with seas; or

3. Graduation from a schoolship approved by and conducted under rules prescribed by Commandant; or

4. Satisfactory completion of a course of training approved by the Commandant and service aboard a training vessel; or

5. Graduation from the U.S. Naval Academy or U.S. Coast Guard Academy

2. Applicant for QMED certificate shall present certificate from U.S. Public Health Service medical officer or other reputable physician attesting that eyesight, hearing, and physical condition are such that he can perform the duties of a QMED.

3. Medical examination is same as for original engineer's license, except that exemption of monocular vision granted to engineers does not apply.

4. Applicants may qualify and receive endorsements for one or more ratings.

SOURCE: Stephen M. Elonka and Joseph F. Robinson, "Standard-Plant Operator's Questions and Answers," Vol. II, McGraw-Hill Book Company, New York, 1959.

SUGGESTED READING _____

COMPLETE STEAM AND REFRIGERATION LICENSE DATA

BOOKS

Elonka, Stephen M., and Joseph F. Robinson: "Standard Plant Operator's Questions and Answers," vol. II, McGraw-Hill Book Company, New York, 1959. (Volume II contains chapter that covers license requirements in more detail than this book. Citizenship, local residence, nonlocal licenses recognized, and fees and renewal for licenses are covered.)

Elonka, Stephen M., and Quaid Minich: "Standard Refrigeration and Air Conditioning Questions and Answers," 2d ed., McGraw-Hill Book Company, New York, 1973. (Contains chapter on license requirements for refrigeration engineers, which are tied in closely with steam licenses where refrigeration compressors are steam driven.)

2

BOILER REGULATIONS, AND HOW TO STUDY

As Chap. 1 shows, the candidate for a fireman's or operating engineer's certificate (license) must meet requirements in age, experience, and character as laid down by local laws and rules. Details can usually be obtained in printed form from the examining boards listed in the previous chapter.

In addition, the applicant must be thoroughly familiar with all laws and regulations covering operation of power plants, particularly those defining qualifications, duties, and responsibilities of the engineer in charge of a plant, and the penalties that may be incurred for neglect.

RULES FOR BOILER CONSTRUCTION: The various codes drawn up by the American Society of Mechanical Engineers (ASME) in connection with the design, construction, and installation of boilers and other pressure vessels have been adopted over practically all of the North American continent. These codes are constantly being revised and kept in line with the latest practice. They provide a service that could not possibly be equaled by any single local organization. Common agreement on a uniform and reliable set of standards is also a tremendous help to designers, manufacturers, and operators.

In addition to these ASME codes, most departments or authorities concerned with the operation of power plants and the licensing of engineers draw up supplementary regulations covering inspection fees, examination requirements, and other items that may have a purely local application.

Greater uniformity in examination requirements and candidates' qualifications for engineers' certificates would be a step in the right direction. The best example is the marine license, which qualifies the holder for a similar berth aboard any ship of similar horsepower in any U.S. port.

ENGINEERING STANDARDS: Matters that do not come within the scope of the codes and other published standards of the ASME are usually covered in the published standards of such other organizations as the American Society for Testing and Materials (ASTM) and the American National Standards Institute (ANSI) in the United States, and the Canadian Engineering Standards Association in Canada.

STUDENTS' EDUCATIONAL BACKGROUND: Three things are essential to proper understanding and answering of examination questions:

1. Sufficient command of English to express ideas clearly after ascertaining the exact meaning of the question.

2. A good working knowledge of elementary mathematical operations.

3. Ability to sketch neatly and to draw in proportion.

The first is needed to answer questions for any grade of certificate; the second, more particularly for higher grades; the third, essential for higher grades but useful for all.

Those preparing for higher grades of certificates must have considerable knowledge of elementary mathematics and mechanical drawing. Mathematics required for lower grades seldom covers more than simple arithmetic, but accuracy in solving even the simplest arithmetical problems is always important. Correctness of all mathematical solutions depends, in the last analysis, on accuracy in addition, subtraction, multiplication, and division.

REFERENCE MAGAZINES AND BOOKS: All operating engineers should subscribe to one or more engineering magazines and have at least a few good textbooks dealing with the various branches of their profession.

The "Standard" series of books listed as references in this book provides the student with a complete home-study course. This book might be considered a primer on boiler room operation. Those wishing to advance will find that "Standard Boiler Operators' Q & A" goes into the subject in greater depth.

"Standard Plant Operators' Manual" has almost 2,000 illustrations, showing step-by-step procedures for such operations as rolling a boiler tube, lining up a pump, and timing a diesel engine, thus providing an "instructor" at the reader's elbow. With many types of instruments installed in modern plants, the two-volume set "Standard Instrumentation Q & A" covers instruments the operator needs to understand. Because electronic instruments (solid-state) are widely used in power plants, the two-volume set "Standard Electronics Q & A" is useful.

"Standard Plant Operators' Q & A" (two volumes) covers everything from boilers and pumps to gas turbines and nuclear reactors. The hydraulics volume introduces the operator to the many hydraulic devices used. These books have been translated and published in some foreign countries, including Japan, Poland, Taiwan, India, and Brazil.

DRAWING INSTRUMENTS AND DRAWING: Ability to make a net freehand sketch and a scale drawing with instruments is a valuable asset to any engineer in his daily work, and is particularly useful in an examination when a sketch or drawing may answer a question better than several pages of writing. Mechanical drawing is the universal language of the engineer.

Very few people have a knack for really good freehand work, but neat sketches, not necessarily to scale, may be made with a few drawing instruments, such as compasses, 60° and 45° triangles, and a 12-in. rule. Most of the sketches for this book were made with drawing instruments, not to scale, but approximately in proportion.

Every progressive engineer should own a complete drafting outfit, including compasses, spring bows, dividers, drawing pens, drawing board, T square, 60° and 45° triangles, protractor, drawing paper, pencils, eraser, and drawing ink, Fig. 2-1.

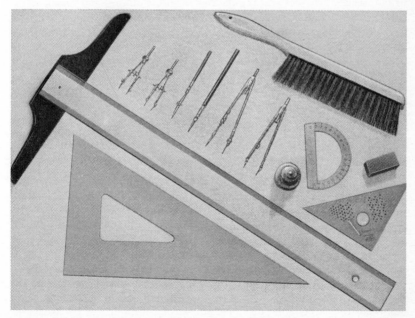

FIG. 2-1 Drawing instruments needed for some examinations.

BOILER ACTS AND RULES GOVERNING EXAMINATIONS: Examiners sometimes ask about laws and rules governing power-plant operation and qualifications and duties of the engineer. Such questions can be answered by quoting the particular section or sections of the act or regulation mentioned in the question. No particular instruction or advice is needed on the method of answering this type of question other than to repeat that a candidate should get copies of all laws and regulations governing these matters and study them carefully, paying special attention to sections dealing with *qualifications* of the engineer, *duties* of the engineer, and *penalties* for neglect of duty or infraction of laws or rules.

An engineer is expected to be familiar with all laws, rules, and regulations governing the conduct of the profession.

QUESTIONS CONCERNING EXPERIENCE: If asked about experience, the candidate should be sure to answer fully. Perhaps the best answer is a direct statement of positions held, in chronological order, plus a few particulars about each plant. Two forms of this type of answer are given on following pages.

ASME CODE AND LOCAL REGULATIONS: Sometimes a candidate is allowed to take copies of the ASME Code and of local regulations into the examination room and to consult them freely. Whether their use is permitted or not, study the codes and regulations carefully beforehand, particularly the sections dealing with use and location of boiler fittings. Use of logarithmic tables, steam tables, and slide rules is permitted at practically all engineering examinations.

AT THE ACTUAL EXAMINATION: Arrange your work neatly. Number each page and leave margins around your writing. Number each answer to correspond with its question and leave spaces between answers. Always put down enough steps in working mathematical problems to explain how you arrived at the result. Write your answers plainly and state units in which answers are given; that is, feet (ft), inches (in.), or pounds per square inch (psi), as the case may be.

TYPICAL QUESTIONS AND ANSWERS

Some of the following questions and answers are on the foregoing topics and others are on routine duties of the engineer. Names used in the answers are, of course, purely imaginary.

Q What are the duties of an engineer in charge of a power plant?
A The engineer sees that his plant is installed so as to comply in every way with regulations governing these matters and that it is operated with due regard to safety and efficiency. He satisfies himself that his assistants are competent to perform their duties and are faithfully following his in-

structions and orders. He draws up rules, and sees that they are carried out, for general machinery adjustment and repair and for all routine operations, such as testing water gages, soot blowing, blowing down and washing out boilers. He draws up a logbook form (see Chap. 20) suitable for his plant and insists that it be an orderly record of all routine operations, repairs, and happenings worthy of note in and around the plant. He personally inspects all important parts, such as interiors of boilers, when they are shut down for cleaning; sees that all repairs and replacements are carried out as soon as reasonably possible; and reports immediately any serious defect to his employer or to the boiler inspector. As he is ultimately held responsible for the plant's safe operation, he uses his own judgment, in the event of any unusual or dangerous situation, as to whether to keep the plant operating or shut down.

Q What penalties may be incurred by an engineer who is guilty of gross neglect of duty, or serious infraction of the laws and rules governing operation of power plants?
A The engineer's certificate may be suspended or canceled. He may also be fined or even imprisoned, in extreme cases, when gross neglect of duty has resulted in grave bodily injury to other persons.

Q What experience have you had as an operating engineer?

NOTE: This question may be answered in either of two ways; (a) by listing jobs in order held, or (b) by an account of experience in narrative form. Following are examples of typical good answers to (a) and (b), using fictitious names.

A (a) LISTING OF JOBS IN ORDER: May 1969 to August 1972. Shift engineer in plant of the Newbury Food Products Co., Newbury, (state or province).
Boiler plant—two 200-boiler-hp horizontal-return-tubular boilers, stoker coal-fired; two 5 × 4 × 6-in. duplex-type boiler feed pumps; Koerting closed-type feed-water heater; two Coffin condensate-return pumps for heating system returns.
Turbine plant—one 75-hp General Electric steam turbine, direct-connected to 50-kW alternator; one old 50-hp horizontal Buckeye noncondensing steam engine (exhaust used for process heating), driving old reciprocating ammonia refrigeration compressor.
August 1972 to present data. Chief engineer in plant of the Standard Mfg. Co., Hillsboro, (state or province).
Boiler plant—two 250-boiler-hp Cleaver-Brooks fire-tube boilers, Bunker C oil fired, one electric-motor-driven Gould centrifugal split-casing feed pump, one 6 × 4 × 10-in. Worthington simplex steam pump for factory water supply, one Bell & Gossett closed feed-water heater.

Turbine plant—One 100-hp DeLaval steam turbine and one 200-hp Elliott steam turbine, direct-connected to 150-kW alternator for supplying current for plant motors and lighting load. One Onan diesel-driven 30-kW emergency electric generator set.

(*b*) EXPERIENCE IN NARRATIVE FORM: I sat for and obtained a second-class engineer's certificate in April 1969, started in May 1969 as shift engineer in the plant of the Newbury Food Products Co., Newbury, (state or province) and continued in this position until August 1974. The boiler plant consisted of [NOTE: Write out list of major equipment as in (*a*)].

In August 1974 I secured my present position as chief engineer with Standard Mfg. Co., Hillsboro, (state or province). [NOTE: Again, detail all the major pieces of equipment as in (*a*).]

In answers like either (*a*) or (*b*) give names of makers of plant equipment, such as boilers, engines, and pumps, where possible so that the examiner knows *exactly* what kinds of equipment you operated.

SUGGESTED READING

Power, 1221 Ave of the Americas, New York, N.Y. 10020.
Steam & Heating Engineer, 35 Red Lion Square, London WC1, England.

3

SAFETY,
DEFINITIONS,
AND FIRE-TUBE BOILERS

The name *fire-tube* derives from the fact that in boilers of this type, all or most of the work is done by heat transfer from hot combustion products flowing inside tubes to the water surrounding them. Such boilers may also be classified as shell boilers — that is, water and steam are contained within a single shell housing the steam-producing elements. Noncylindrical sections and flat surfaces are given added resistance to internal pressure by various means: diagonal stays, through-bolts, or tubes which are flared at the tube sheets to act as stays.

In such a shell, the force tending to burst it along the length is twice that tending to burst it around the girth. Thus high pressures and large diameters would lead to extremely thick shell plates. Hence there is a definite economic limit on the pressure and capacity that can be reached with shell-type boilers.

An operating pressure of 250 psi (pounds per square inch) may be considered the practical ceiling, and in the United States, capacity rarely exceeds 25,000 lb of steam per hr — roughly 750 boiler horsepower. In Europe, where larger fire-tube boilers have always been popular and economical, and boiler-code conditions are different, units of 30,000 lb per hr are not unusual.

As a class, fire-tube boilers feature simple and rugged construction and relatively low first cost. Their characteristically large water capacity makes them somewhat slow in coming up to operating pressure, but provides some steam accumulator action that makes it possible to meet load changes quickly.

While horizontal-return-tubular and locomotive-type boilers are built today only on special order, there are a number still in use, and examiners still ask questions on their construction and operation.

In this chapter we also discuss boiler safety.

SAFETY METHODS

Q What are the objects of the various boiler acts and regulations?

A Main objects are public safety and efficiency. Principal means of securing these are:

1. Setting up high standards for materials, design, and workmanship in the construction of steam boilers and pressure vessels to ensure strength, durability, and safety.

2. Provision for periodic inspection of boilers and pressure vessels by qualified inspectors appointed for that purpose.

3. Certification of properly qualified engineers to have charge of and operate steam power, process, and heating plants.

Q How would you get a boiler ready for inspection?

A Cool the boiler slowly. When it is cool, open the blowoff valve and drain the boiler. At the same time, open some small valve above the water line to permit air to enter the boiler as it drains and so prevent formation of a vacuum. When boiler is drained, remove manhole and handhole cover plates, wash out the interior, and remove any loose scale and mud. Clean soot from outside of tubes and shell and inside of firebox. Remove all soot and ashes from ashpit and combustion chamber. Have ready all tools, clamps, and fittings that may be needed for valve or gage setting or testing. If the boiler is in a battery with others, make sure that steam, water, and blowoff valves cannot be accidentally opened while men are working in the empty boiler.

Q How would you assist the boiler inspector during his inspection?

A Give him all the help he requires. Point out any known defects. Station someone immediately outside the boiler when he is making the internal inspection and make sure, if the boiler is in a battery with others, that all steam, water, and blowoff valves are locked shut or otherwise secured, so that they cannot be opened accidentally. Make provision for the application of a hydrostatic test, if this is required, and, in general, assist in every way to make the inspector's examination thorough and complete.

Q How is a hydrostatic test applied?

A Fill the boiler until water comes out of the air vent, then close all valves, gag the safety valve so that it cannot open to release the pressure, and apply and hold water pressure of $1\frac{1}{2}$ times the maximum allowable working pressure, by means of a boiler-feed pump or a hand pump if other pumps or sources of water pressure are not available. While the water pressure is on, the outside of the boiler is examined for leaks, distortion of plates, or other defects, and light hammer blows may be applied to test the strength of any parts suspected of weakness.

Q What action would you take on discovering a dangerous defect in any boiler or engine under your charge?

A I would shut down the boiler without delay and immediately notify the local boiler inspecting authority and my employer.

If I discovered a dangerous defect in an engine under my charge, I would immediately shut down the engine and take steps to replace or repair the defective part.

Q What should be your first duty on taking over a shift?

A Check the water level in the boilers or have the firemen or water tenders check it under my supervision, by blowing down the water gage glasses and the water columns if the boilers are fitted with water columns. I would note if the water returned quickly to its proper level in the glass when the drain valve was closed, thus showing the passages to be free from obstruction, and I would check the level in each glass with the gage cocks.

Q After the water level in the boilers has been checked, what should be the next procedure?

A Check the steam pressure, then examine the logbook for anything out of the ordinary routine on the previous shift and for any instructions left for the oncoming shift. After water levels have been checked, a complete inspection should be made of the entire plant. Examine all lubricators and oil pumps to see that they are full and working properly. Feel accessible bearings for signs of excessive heating. Read all recording gages, meters, and switchboard instruments and note any deviation from normal. Ascertain cause of any unusual noise, and remove or remedy it. Make this plant inspection immediately on taking over, as the time to find trouble left over from the previous shift is at the start of a shift and not some hours later.

Q What certificates (licenses) should be displayed in the plant engine room?

A Certificates of the engineers operating the plant, boiler and pressure vessel inspection certificates, and any others that may be required by local authorities.

BOILER DEFINITIONS _____

Q What is a steam boiler (also known as a steam generator)?
A It is a closed vessel, strongly constructed of steel or iron. When in use it is partly filled with water, which is converted into steam by the external application of heat. The steam thus generated is used for power, heating, or manufacturing purposes.

Q What is meant by steam space in connection with steam boilers?
A A steam boiler is only partly filled with water when in operation. Remaining space is called steam space because it is needed for the disengagement of steam from the water and for storage of this steam until it is drawn off through the steam main.

Q What constitutes the heating surface of a steam boiler?
A The heating surface consists of all parts of the boiler that have water on one side and fire or hot gases on the other. When the boiler is fired, there is a continual transfer of heat from the fire or hot gases on the one side, through the plate, to the water on the other side.

Q What is grate surface?
A This is the area of the grate upon which fire rests, in a coal or wood-fired boiler. It is usually measured in square feet. Thus the area of a fire grate 6 ft long by 5 ft wide would be $6 \times 5 = 30$ sq ft.

Q In connection with steam boilers, what is meant by (1) *water line*, (2) *fire line*?
A 1. The *water line* is the level at which the water stands in the boiler. This level should always be higher than any part of the boiler that is exposed to excessive heat and likely to be damaged if not covered by water.
  2. The *fire line* is the highest point of the heating surface in most common types of boilers, but this definition cannot be applied to all boilers. The fire line is level with the top row of tubes in the horizontal-return-tubular and dry-back marine boilers, at the highest point of the crown sheet in the locomotive boiler, and at the upper tube sheet in the wet-top vertical fire-tube boiler. The fusible plug is so located in each of these boilers that it will give warning of low water before the water level falls below the fire line.
  In the dry-top vertical fire-tube boiler and many types of water-tube boiler, there is no definite line at which water-heating surface ends, and the bottom of the gage glass in these boilers simply indicates the point below which it is deemed unsafe to carry the water level.

Q What are some of the requirements of a good type of steam boiler?
A 1. Strong and simple construction.
  2. Materials and workmanship of highest standard.

3. Design that ensures constant circulation of water in boiler, thus distributing heat evenly through the entire body of water and keeping the various parts of heating surface as nearly as possible at the same temperature.

4. Large area of heating surface to ensure the utmost possible transfer of heat from the hot gases in the furnace to the water in the boiler.

5. All parts of boiler readily accessible for repair, inspection, and cleaning.

6. Ample combustion space so that gases will be completely burned before passing to the chimney.

7. Large steam space so that steam is able to rise freely from surface of water.

Q How are boilers classified?

A Boilers may be classified in the following ways: (1) according to direction of axis of shell, *vertical* or *horizontal;* (2) according to use to which they are put, *stationary, portable, tractor,* or *marine;* (3) according to location of furnace, *internally fired* or *externally fired;* (4) according to relative positions of water and hot gases, *water-tube* or *fire-tube;* (5) as special boilers — *electric, once-through,* etc.

Q Describe the horizontal-return-tubular boiler according to classifications in the preceding question.

A Horizontal-return-tubular boiler, Fig. 3-1, commonly called *hrt,* is a horizontal, stationary, externally fired fire-tube boiler.

Q What is the difference between a tube and a flue?

A Both are cylindrical tubes, but the term *tube* is usually applied to those of small diameter, up to about 6 in. Over this diameter, they are called flues. Note that tube sizes refer to *outside* diameters.

Q What is the difference between a fire tube and a water tube?

A A fire tube has water on the outside of the tube, while fire or hot gases pass inside the tube. A water tube has the opposite condition; that is, water passes through the inside and fire or hot gases are on the outside.

FIRE-TUBE TYPES

Q Describe the vertical dry-top boiler.

A This is a vertical, internally fired fire-tube boiler. As shown in Fig. 3-2, it consists of a vertical cylindrical shell containing a cylindrical firebox and a number of small fire tubes. Heat radiated from the fire passes through the firebox plates to the water in the boiler. The hot gases pass upward through the fire tubes to the smokestack, giving up part of their

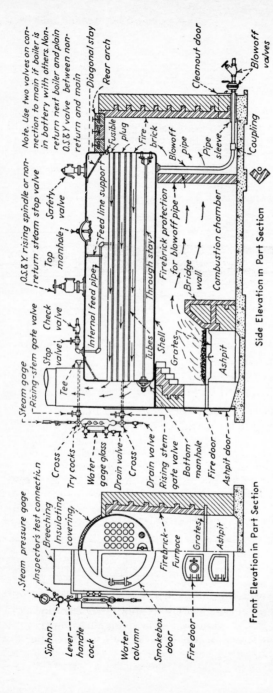

FIG. 3-1 Horizontal-return-tubular boiler (hrt) is an older design but still found in many plants.

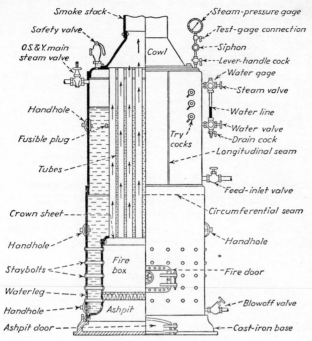

Smoke stack

Safety valve

O.S.&Y. main
steam valve

Cowl

Steam-pressure gage

Test-gage connection

Siphon

Lever-handle cock

Water gage

Steam valve

Water line

Water valve

Drain cock

Longitudinal seam

Handhole

Fusible plug

Tubes

Try
cocks

Crown sheet

Handhole

Staybolts

Waterleg

Handhole

Ashpit door

Fire
box

Ashpit

Feed-inlet valve

Circumferential seam

Handhole

Fire door

Blowoff valve

Cast-iron base

FIG. 3-2 Dry-top vertical tubular fire-tube boiler.

heat to the metal of the tubes, which in turn transfer the heat to the water in the boiler.

The upper tube plate forms the top head of the boiler. As the water level is carried about two-thirds of the length of the tubes above the lower tube sheet, or crown sheet, the top tube sheet is dry, that is, it has steam on one side and hot gases on the other side, hence the name *dry top*. The main drawback in this form of construction is the possibility of the tubes becoming overheated above the water line and loosening in the tube sheet. The fact that the upper third of the tubes is dry gives a slight amount of superheat to the steam.

Handholes and cleanout plugs are provided at convenient points for washing out and inspection. In the smaller sizes, the bottom of the shell rests on a cast iron base that also forms the ashpit. Small vertical fire-tube boilers are used principally for driving portable machinery. Larger sizes, up to several hundred horsepower, are sometimes installed in stationary plants where floor space is limited. Pressures seldom exceed 200 psi.

Q Describe the vertical wet-top, or submerged-top, boiler.

A This boiler, Fig. 3-3, is similar in construction to the dry-top vertical except for the top head. In the wet-top boiler, a conical-shaped plate is riveted to the top head, and the upper tube sheet forms the bottom of this cone. This enables the water line to be carried above the upper tube sheet so that it and all the tubes are below the water level: hence the name *submerged-top* or *wet-top* boiler. Other construction details are the same as in the dry-top boiler.

Q What are the advantages and disadvantages of vertical fire-tube boilers?

A The advantages are: (1) compactness and portability; (2) low first cost; (3) very little floor space required per boiler horsepower; (4) no special setting required; (5) quick and easy installation.

The disadvantages are: (1) the interior is not readily accessible for cleaning, inspection, and repair; (2) water capacity is small, making it difficult to keep a steady steam pressure under varying load; (3) boiler is liable to prime (pull over water with the steam) when under heavy load because of the small steam space; (4) efficiency is low in the smaller sizes

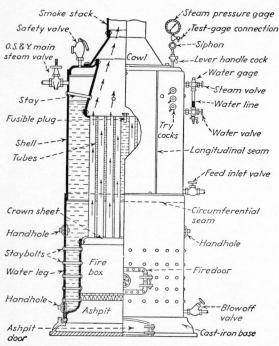

FIG. 3-3 Wet-top vertical tubular fire-tube boiler.

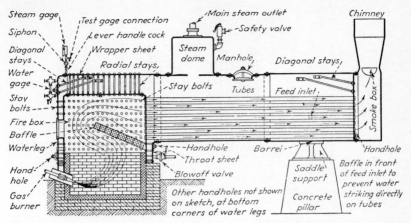

FIG. 3-4 Locomotive-type boiler is older firebox or water-leg type seldom found in use today.

(the hot gases have a very short and direct path to the stack, and much of the heat goes to waste).

Q Describe the locomotive-type boiler, and name all its parts.

A This type, Fig. 3-4, was used on locomotives and steam tractors, also for portable and stationary work. Like the vertical-tubular, the loco-motive-type boiler is an internally fired fire-tube unit, but its *shell* is horizontal and the *firebox* is not contained within the cylindrical portion of the boiler. The firebox is rectangular in shape with a curved top or *crown sheet*. This crown sheet is supported by *radial stays* screwed into the crown sheet and the outer *wrapper sheet* and riveted over.

The *inner side sheets* of the firebox are connected to the *outer side sheets* by short screwed stays called *stay bolts*. Space between these sheets is called a *water leg*. The *fire tubes* are within the barrel and run from the *firebox tube sheet* to the other head of the barrel, this head being in a *smokebox* formed by extending the barrel beyond the *tube sheet*.

The *firebox front sheet*, above the crown sheet, and the *smokebox tube sheet*, above the tubes, are supported by *longitudinal stays*. In some cases *diagonal stays* also are used for this purpose.

The *steam dome* provides additional steam storage space and allows the *main steam outlet* to be taken off at a considerable height above the water line, thus reducing possibility of water being carried over with steam.

Q Describe the dry-back marine-type boiler.

A This boiler, Fig. 3-5, is an adaptation to stationary practice of the well-known Scotch marine wet-back boiler. It consists of an outer cylindrical

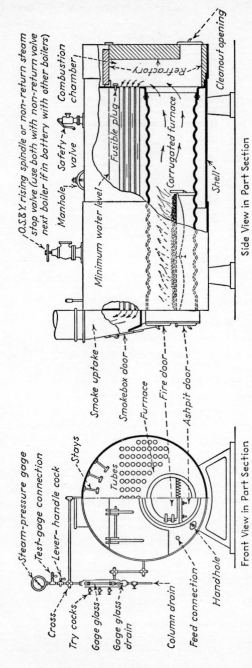

FIG. 3-5 Dry-back Scotch marine-type boiler has two gas passes and corrugated furnace. Today this boiler comes "packaged" and fully automated; it is widely used.

shell, of short length but large diameter, and contains one or two large furnace flues of corrugated pattern, the better to withstand expansion and contraction stresses. The hot gases from these furnaces pass into a brick combustion chamber at the back of the boiler and return through the fire tubes to the front of the boiler and thence to the chimney.

In the wet-back marine boiler, the shell, tube, and furnace construction are similar, but the combustion chamber is contained within the shell, and no outside setting is required. The dry-back type is a quick steamer because of its large heating surface, is compact and easily set up, and shows fairly good economy.

Q Describe the hrt boiler.

A This boiler, Fig. 3-1, has been the most common form of externally fired fire-tube boiler and is still used in small- or medium-sized power plants. Unlike the vertical-tubular, locomotive, or dry-back marine-type boilers, the hrt contains no firebox and must be externally fired. This makes a brick setting necessary to provide furnace and combustion space. The boiler itself is a cylindrical shell with flat ends or heads and contains a large number of fire tubes.

The grates are beneath the front part of the boiler shell. The hot gases from the furnace pass along the bottom of the shell to the rear, returning through the fire tubes to the smokebox at the front and thence to the stack. This *return*-tube arrangement gives the hot gases a much longer travel than they have in the *direct*-tube arrangement of the vertical-tubular or locomotive boilers, where they pass directly from the fire through the tubes to the chimney.

Small hrt boilers are often supported by brackets riveted or welded to the shell and resting on the brick side walls of the setting. Large hrt boilers are always suspended by suitable bolts and lugs from steel cross beams supported on steel columns.

Q Describe a three-pass firebox design boiler.

A In the three-pass category we find a number of designs that are basically firebox boilers as distinguished from the internal-furnace type (Scotch-type, Fig. 3-5) with its cylindrical furnace tube. Typical units have a firebox with a flat bottom and vertical sides going into an arched crown sheet that forms the top, Fig. 3-6.

This furnace shape requires stays, as do the water legs at the sides of the rear chamber into which the firebox discharges. Here the aim of the designer is for a large combustion space and a higher ratio of furnace to tubular surface.

Q What is a "packaged" fire-tube boiler?

A Packaged boilers represent the bulk of fire-tube boilers manufactured today. But water-tube designs also come packaged. The American Boiler Manufacturers' Association defines a packaged fire-tube boiler as a "modified Scotch-type boiler unit, engineered, built, fire-tested before shipment, and guaranteed in material, workmanship and performance by one firm, with one manufacturer furnishing and assuming responsibility for all components in the assembled unit, such as burner, boiler, controls and all auxiliaries. . . ."

Packaged boilers burn solid fuel as well as oil or gas, or a combination of these fuels. In the two-pass fire-tube design, Fig. 3-5, furnace gas reverses direction at the rear head, then travels back through the tubes to the smoke outlet. In the three-pass unit, Fig. 3-7, gas reverses direction three times at rear and front heads. A four-pass design is also manufactured, adding yet another gas-flow reversal.

These internal-furance boilers of the so-called Scotch type are essentially self-contained. Combustion takes place in a cylindrical furnace within the boiler shell. Fire tubes run the length of the shell at the sides of, and above, the internal furnace, which may be straight as in Fig. 3-7 or corrugated as in Fig. 3-5. Gas from the furnace reverses direction in a chamber at the rear and travels forward through the tubes, where it may again reverse, depending on the number of passes, finally exhausting to a smokebox and up the flue.

Figure 3-5 shows the dry-back design, with a refractory-lined rear chamber. When the rear chamber is submerged in the water space (wetback design), we have the distinctive Scotch marine boiler.

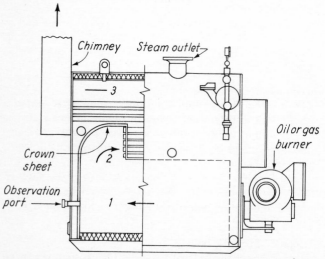

FIG. 3-6   Firebox design has arched crown sheet, three gas passes.

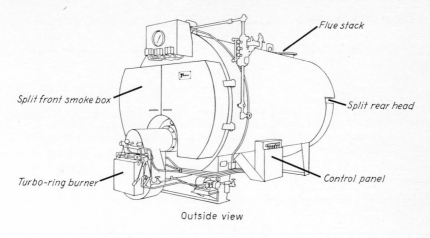

Outside view

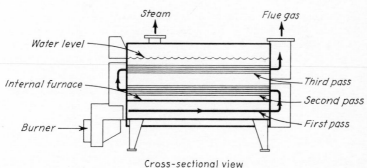

Cross-sectional view

FIG. 3-7 Modern dry-back Scotch marine type has three gas passes, comes complete with fittings and ready to hook up to steam outlet, water and/or gas and electricity.

The internal furnace is subject to compressive forces and so must be designed to resist them. Thus furnaces of relatively small diameter and short length may be self-supporting if wall thickness is adequate. For larger furnaces, one of four methods of support may be used: (1) corrugating the furnace walls, (2) dividing the furnace length into section with a stiffening flange (Adamson ring) between sections, (3) using welded stiffening rings, and (4) installing staybolts between the furnace and the outer shell. If solid fuel is to be fired, a bridge wall may be built into the furnace at the end of the grate secion.

The outside view of the unit in Fig. 3-7 shows the turbo-ring burner, control panel, split rear head and unit ready to be slid into the boiler room and hooked up to water, fuel, and electric connections.

SUGGESTED READING

REPRINT SOLD BY *Power* MAGAZINE,
Steam Generation, 48 pp.

BOOKS
   "ASME Boiler and Pressure Vessel Code," sections 1–4, American Society of
   Mechanical Engineers, New York, 1974.
   Elonka, Stephen M., and Anthony L. Kohan: "Standard Boiler Operators' Ques-
   tions and Answers," McGraw-Hill Book Company, New York, 1969.
   Elonka, Stephen M.: "Standard Plant Operators' Manual," 2d ed., McGraw-Hill
   Book Company, New York, 1975.

4

WATER-TUBE
AND SPECIAL BOILERS

In contrast to the fire-tube idea, water-tube designs feature one or more relatively small drums with many tubes, in which a water-steam mixture circulates. Heat flows from the outside of the tubes to this inside mixture. Because of this subdivision to pressure parts, larger capacities and higher pressures are possible — 5,000 psi in some supercritical designs.

Today, modern shop-assembled water-tube boilers have larger headers and drums for better steam quality and ready access for inspection, improved design of small steam separators, more radiant heating surface, increased gas travel, fully drainable superheaters, better protection against corrosion so the unit can be used for fuels of higher sulfur content, and many other design improvements.

WATER-TUBE BOILERS _____

Q Describe a small-sized longitudinal-drum water-tube boiler.
A Figure 4-1 shows an early small-sized longitudinal-drum boiler with one drum; larger boilers have two or three drums. Drums run from front to rear of boiler. The inclined straight steel tubes, usually around 4 in. outside diameter, are connected with the drum by pressed-steel headers. A mud drum below the rear headers collects sediment and is blown out from time to time. Tube headers are in one piece for each vertical row of tubes. Staggered tube holes force hot gas to follow a tortuous path as it

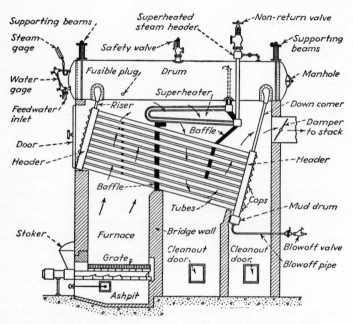

FIG. 4-1 Longitudinal-drum water-tube boiler is older design, has baffles for directing hot gases through three passes.

passes up around the water tubes. Header handholes, for tube cleaning, are closed by bolted covers with machined joints. The superheater, usually fitted in the space below the steam drum, is a set of U tubes. Saturated steam from the drum passes downwards to be superheated in these tubes, then passes from the bottom superheater header to the main stop valve and header.

The boiler is suspended from cross beams attached to the steam drum and supported by steel columns.

Q Have water-tube boilers any advantages over fire-tube types?
A The water-tube boiler is supposedly safer, largely because the bulk of water is in small units—the tubes. If a tube ruptures, only a comparatively small body of water is released to flash into steam instantly. As a rule, all parts of the water-tube boiler are readily accessible for cleaning, inspection, and repairs. Water-tube boilers are faster steamers because of their large heating surface, long gas travel, and rapid and positive water circulation. For the same reasons, they can carry much greater overloads and respond more readily to sudden changes and fluctuations in demand.

Q Describe the Stirling water-tube boiler.
A This is a bent-tube type of water-tube boiler. There are now many variations and adaptations, but the original standard Stirling boiler has three upper steam drums, connected by curved tubes to a single mud drum at the bottom. Steam drums are connected to each other by short curved tubes, as in Fig. 4-2. These equalize steam pressure and water level in the three drums, though water level usually varies a little in operation. Hot furnace gas is guided between the water tubes by baffle tiles behind the tube banks. Feed water enters the rear drum, moves down the rear bank of tubes, then up the front banks. Steam is delivered from the top center drum. Upper drums are suspended from steel cross beams bolted to steel columns. The lower mud drum hangs freely by the tubes from the upper drums, and thus can move in any direction as the boiler parts expand and contract with temperature changes. The furnace is enclosed in a brick setting faced with a refractory material and is usually air-cooled or water-cooled.

Q Describe some form of vertical straight-tube water-tube boiler.
A The vertical water-tube boiler, Fig. 4-3, consists of two cylindrical drums set with their axes vertical and connected by a large number of vertical water tubes. The boiler is enclosed in a brick setting and supported by brackets fastened to the lower drum and resting on steel or concrete pillars. This construction allows the boiler to expand and contract freely without disturbing the setting to any great extent. A vertical tile baffle is placed in the center of the tube bank to divert the products of

combustion from the furnace upward through the front bank of tubes and then downward through the rear bank. The water circulation is in the same direction, up the front tubes and down the rear tubes. A sectional view through the boiler setting is shown in Fig. 4-3.

Q What are modern trends in steam boiler design and construction?
A Main changes are in the smaller ones, such as the dry-back marine fire-tube type, which today is shop-assembled as a packaged boiler. They come complete with automatic controls.

Water-tube boilers are also packaged today, in sizes up to 325,000 lb per hr. The coil-type boiler, similar to the once-through, Fig. 4-9, is another newer type. In general, automatically fired packaged units have in recent years come into such prominence that the day is almost here when a smaller hand-fired, field-erected boiler will be considered a museum piece, along with the steam engine.

Welding has replaced riveted shells and drums. Nozzles are welded to

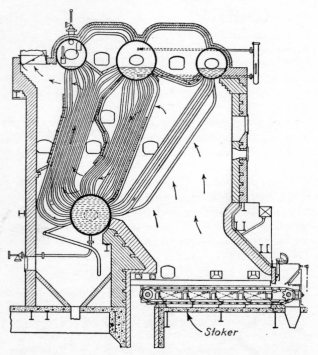

FIG. 4-2 Stirling water-tube boiler has stoker for automatic firing of coal; it is an older design but still in use.

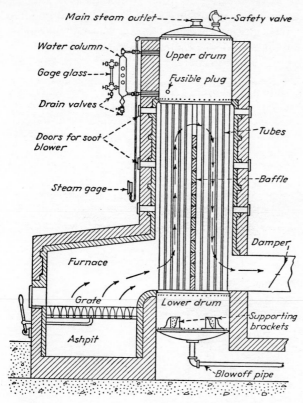

Main steam outlet - - - - → ⟍ ←- - Safety valve

Water column ⟍

Upper drum

Gage glass - - -

Fusible plug

Drain valves ⟨

Doors for soot ⟨
blower

Tubes

Steam gage - - - →

Baffle

Damper

Furnace

Grate

Lower drum

Supporting
brackets

Ashpit

Blowoff pipe

FIG. 4-3 Vertical water-tube type saves floor space, is older design.

shells and many minor details that formerly involved riveting are now done by electric or oxyacetylene welding.

Pressures have increased little on fire-tube boilers, chiefly because thickness of firebox plates and of shell plates directly exposed to intense furnace heat cannot be increased without danger of burning the plates on the fire side. With a very thick plate, heat cannot pass through rapidly enough from fire side to water side to prevent overheating.

Water-tube boilers are invading the small boiler field in increasing numbers. There are two main types: the two-drum boiler, with an upper and a lower drum connected by bent tubes, and the three-drum boiler, with one upper drum and two lower drums connected by bent tubes.

Small packaged water-tube boilers of the coil type, Fig. 4-7, are today widely used because they are comparatively light and quick-steaming.

These designs have small, generally coiled steel tubes, with forced circulation and intense heat release from a gas or oil burner. They usually come to full capacity and pressure within 5 minutes after start-up. Operation is automatic. Some produce as low as 1,500 lb per hr steam, others produce 15,000 lb per hr at 900 psi.

Large modern boilers are practically all of the bent-tube water-tube multidrum design. This very flexible form of construction lends itself readily to a great variety of tube and drum arrangements. Drum diameters are small; heavy drums are of all-welded construction. Steam pressures have climbed up to 5,000 psi in supercritical designs.

Furnace walls are air- or water-cooled. Preheating air for combustion by passing it through flue-gas-heated air heaters utilizes heat that would otherwise be wasted. Water walls protect the setting from furnace heat

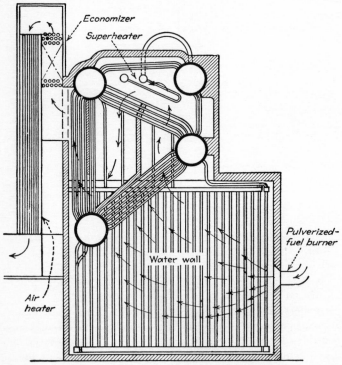

FIG. 4-4 Water-tube boiler with waterwall around furnace has superheater, economizer, and air heater.

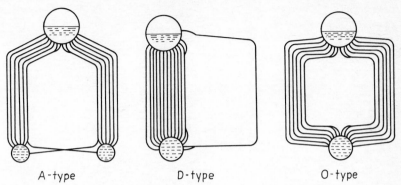

A-type   D-type   O-type

FIG. 4-5 Medium-sized water-tube boilers today come in these three general designs.

and form part of the boiler itself, since water from the drums circulates through the wall tubes and absorbs radiant furnace heat. Crushing the coal and firing it in powdered form is also common practice in large boiler plants.

Q Are water-tube boilers of the packaged type available for burning coal?

A Yes, at least one coal-fired packaged boiler today is fully automatic, with a rating of only 10,000 lb of steam per hr. This water-tube unit burns bituminous coal on a pulsating-grate stoker. The furnace is cooled with water-jacketed walls, thus reducing furnace-wall insulation. Provision is made for ash removal by a screw conveyor into ash cans. Banked fire can be maintained and normal operation resumed without any need for human intervention. This boiler is equipped with both forced-draft and induced-draft fans. As in the boiler shown in Fig. 4-1, the tube bundle is of the inclined straight-tube type, with the longitudinal steam drum above.

Q Sketch a large, modern water-tube boiler with waterwall furnace, air preheater, superheater, economizer, and pulverized-fuel system of firing.

A Figure 4-4 shows an outline sketch of a large water-tube boiler and furnace of modern design.

Q Describe a modern packaged water-tube boiler.

A Modern shop-assembled water-tube boilers today come in the three basic tube and drum arrangements shown in Fig. 4-5. They are known as the *A type, D type,* and *O type.*

Figure 4-6 shows the complete assembly of an O-type arrangement as shipped from the manufacturer. It consists of convection tubes, wa-

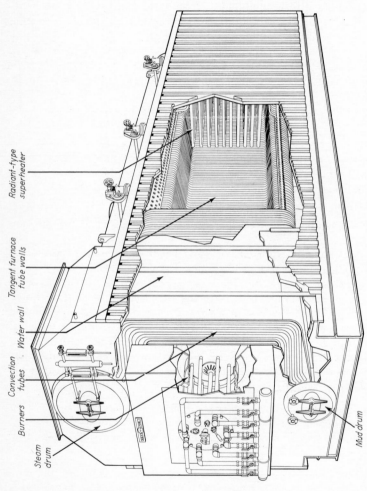

Radiant-type
superheater

Tangent furnace
tube walls

Convection
tubes · Water wall

Burners

Steam
drum

Mud drum

FIG. 4-6 Modern water-tube boiler is packaged, shipped from manufacturer as shown, and ready to slide into position and hook up. The costly field-erected setting needed for older units is no longer required.

terwalls, tangent furnace tube walls and a radiant-type superheater. The unit is furnished with burner systems for burning various kinds of fuel.

These packaged units are designed to produce from 6,000 to 350,000 lbs of steam per hr, and more. They come rated for steam pressures up to 950 psi. The generator tubes and water-cooled walls direct the flow of hot gases from the front of the unit through the furnace and around both sides at the rear, then into convection zones and toward the front, where the flue gas is discharged up the stack.

Because the manufacturer shop-assembles these units, then ships to customers, field-erected boilers, except for large utility plants, are a thing of the past. The builder takes responsibility for the entire unit.

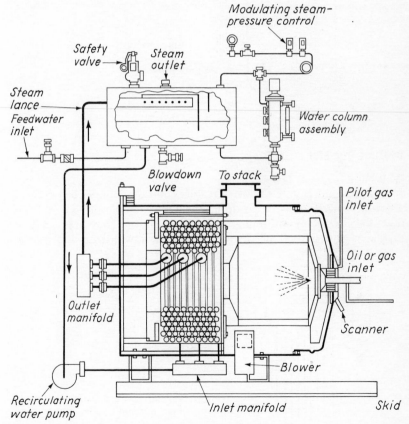

FIG. 4-7 Multiple coil water-tube boiler is fast steaming, has forced circulation and drum for water-steam separation.

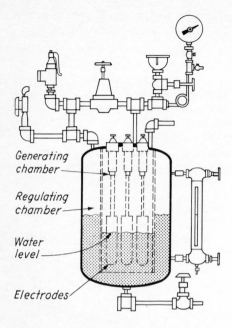

FIG. 4-8 Electric boiler stops steaming when water level drops.

SPECIAL BOILERS

Q What is a special boiler?
A Any design which has unusual features and thus cannot be labeled as a fire-tube, water-tube, or cast-iron type. One example is the coiled tube unit, Fig. 4-7. Rapid steaming ability is a feature. They are small (for capacity), generally of coiled steel tube bundles, with forced circulation

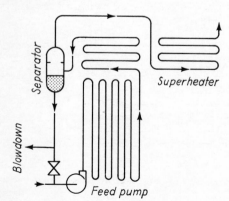

FIG. 4-9 Once-through design is one continuous tube.

and intense heat release from gas or oil burner. They often come to full capacity and pressure within 5 minutes after cold startup, and operation is automatic.

Q What is an electric boiler?
A Figure 4-8 is an electric unit in which the resistance of water between solid metal electrodes generates heat. Obviously not a fire-tube type, it is really a vertical shell-type design. Protection against overheating in case of low water level is automatic, since electric current flows only through the water. If there is no water, there is no current flow, making it one of the safest boilers in case of low water. Chemical treatment improves the conductivity of the water for faster steaming.

Q Explain supercritical pressure.
A At the critical pressure of 3,206.2 psia (pounds per square inch absolute), steam and water coexist at the same density. When this high-pressure mixture is heated above its 705.4°F saturation temperature, dry superheated steam is produced. Steam generators used for this service are the once-through type, with feedwater being pumped in at one end and dry steam coming out the other, thus no steam drum is needed. Such units are used only in large public utility plants.

Q Explain the once-through boiler.
A Figure 4-9 is a schematic diagram of the once-through design, which has only *one* single tube, into which goes feedwater at one end and out of which comes saturated, or superheated steam at the other end. At pressures below critical (3,206.2 psi) a once-through unit may have a separator to deliver saturated steam to the superheater, as in Fig. 4-9, and to return the collected moisture to the feed-pump suction as shown.

The once-through cycle is ideally suited for pressures above the critical point where water turns to steam without actually boiling. Another advantage is that the heavy and costly steam drum is not needed.

SUGGESTED READING

REPRINT SOLD BY *Power* MAGAZINE,
    Steam Generation, 48 pp.

BOOKS
    Elonka, Stephen M., and Anthony L. Kohan: "Standard Boiler Operators' Questions and Answers," McGraw-Hill Book Company, New York, 1969.
    Elonka, Stephen M., and Joseph F. Robinson: "Standard Plant Operators' Questions and Answers," vols. I and II, McGraw-Hill Book Company, New York, 1959.
    Elonka, Stephen M.: "Standard Plant Operators' Manual," 2d ed., McGraw-Hill Book Company, New York, 1975.

5

CONSTRUCTION MATERIALS, WELDING, AND CALCULATIONS

In addition to the materials covered in this chapter, metals for higher pressures and temperatures are being developed. Such high-quality alloys must contain expensive alloys for the superheater and reheater sections of large, modern water-tube steam generators.

Today, low-carbon steel is used in most water-tube boilers operating over a wide temperature range. Convection temperatures run between 500 and 700°F. Medium-carbon steel, with 0.35 percent maximum of carbon, permits higher stress levels than low-carbon steel at temperatures up to 950°F, for example. But for superheater tubes, which must resist temperatures above 950°F, alloy steels are required. These may contain chromium, chromium-molybdenum, and chromium-nickel.

Although repairs on the pressure components of boilers must be made only by certified welders, we cover a few questions on welding and calculations that every operator should know.

CONSTRUCTION MATERIALS _____

Q List the principal materials used in steam-boiler construction and name some of the parts for which each is especially suitable.

A WROUGHT STEEL (also called *low-carbon* steel): For boiler plates, bolts, tubes, rivets, stays, pipes, reinforcing rings, manhole door frames, mud drums, nozzles, manhole and handhole covers.

CAST STEEL: For high-pressure fittings, valves, tube headers, manhole and handhole cover plates, supporting lugs.

WROUGHT IRON: For tubes, pipes, stay bolts, rivets, door-frame rings.

CAST IRON: For valves, pipes, fittings, and water columns when pressures do not exceed 250 psi and temperatures are not over 450°F. Cast-iron fittings on blowdown lines are limited to 100 psi.

MALLEABLE CAST IRON: For pipe fittings, water columns, valve bodies.

BRASS: For small valves, gage glass fittings, parts of gages, pipe.

BRONZE: For safety-valve seats and same purposes as brass.

COPPER: For plates and tubes.

ALLOY STEELS (steels containing small percentages of other metals, such as nickel, chromium, molybdenum): Commonly used when greater strength is required and metal is subjected to high temperatures.

Q Explain briefly how wrought steel, cast steel, wrought iron, cast iron, malleable cast iron, chilled cast iron, brass, bronze, copper, and alloy steels are manufactured.

A WROUGHT STEEL: Composed primarily of the element iron, combined with small percentages of other elements, notably carbon (generally less than 0.3 percent). First step in the manufacture of iron and steel is the smelting of iron ore in a blast furnace, with limestone as a flux and coke as fuel. The molten metal is cast in narrow molds as "pigs." The process of manufacturing wrought steel consists of melting this pig iron again in a suitable furnace, usually an "open hearth," burning off the carbon and other impurities, then adding whatever amount of carbon may be required for the desired grade of steel. The finished steel usually contains small percentages of manganese, sulfur, and phosphorus as well as carbon. Molten metal from the steel furnace is run off into ingot molds, and these ingots are rolled, forged, or pressed into plates, bars, angles, and other shapes.

CAST STEEL: Cast directly from the steel furnace into molds of any desired shape and not formed by rolling or any other process.

WROUGHT IRON: Made by melting pig iron in a "puddling" furnace and working it in such a way as to remove practically all carbon. The pasty mess that is left is almost pure iron.

CAST IRON: Made by melting pig and scrap iron in a cupola furnace, then running the molten metal into molds. Iron castings made in this way may contain a large amount of free carbon; that is, carbon that is not combined with the iron but is in the form of graphite. Cast iron high in free carbon is called gray cast iron because of its gray appearance when machined. It machines easily and is rather weak in structure. Cast iron that is low in free carbon is somewhat more like steel. It has a white appearance when machined and is harder and stronger than gray cast iron.

MALLEABLE CAST IRON: Produced by annealing "white-iron" castings. Packed with oxide or iron in cast-iron boxes, the castings are heated to red heat. They are kept at this temperature for a considerable time, then allowed to cool slowly. This makes the castings more or less malleable — that is, capable of being bent or worked to some extent.

CHILLED CASTINGS: Made in special molds that cool or chill the outer surface of the casting rapidly, thus making it very hard.

COPPER: Copper ore is first crushed fine, then washed, screened and concentrated, rewashed, and smelted to a *copper matte.* The matte goes through a further refining process to burn out such impurities as sulfur and iron. The resulting *blister copper,* about 99 percent pure, is further refined before being manufactured into wire, rods, and bars.

BRASS: Principal ingredients are copper and zinc. Some brasses contain small amounts of tin and lead as well. They can be made soft or hard by varying the proportions and the amount of cold working.

BRONZE: Contains copper with either zinc or tin, or both. Many modern bronzes contain such elements as aluminum and nickel.

ALLOY STEELS: Made by the same processes as wrought and cast steel, but containing small percentages of other metals, such as chromium, nickel, and manganese. In boiler work, alloy steels are used for plates and castings for high temperatures (750 to 1100°F).

Q The following eight terms refer to the physical properties of materials. Explain their meaning briefly: ductility, elasticity, malleability, hardness, toughness, homogeneity, tenacity, resilience.

A 1. DUCTILITY: Ability to withstand drawing out or other deformation without breaking.

2. ELASTICITY: Ability to return to normal shape after a deforming force has been removed.

3. MALLEABILITY: Capable of having the shape changed by rolling, hammering, or bending, without cracking or breaking.

4. HARDNESS: Ability to withstand surface wear or abrasion.

5. TOUGHNESS: Measure of a material's ability to stand up without fracture under repeated twisting or bending.

6. HOMOGENEITY: Refers to the internal structure of a material. When broken, homogeneous material shows uniform grain or fiber.

7. TENACITY (high tensile strength): Ability to stand large pulling force.

8. RESILIENCE: Ability to store up energy under stress and give it back when stress is removed.

PHYSICAL PROPERTIES

Q What are the physical properties of low-carbon steel, wrought iron, cast iron, cast steel, and brass?

A Only a very general answer can be given, since all these metals vary widely in quality because of different manufacturing methods, varying proportions of alloying metals, and impurities that cannot be entirely eliminated. Broadly speaking:

Low-carbon steel is ductile, malleable, tenacious, tough, elastic, and fairly homogeneous.

Wrought iron possesses much the same physical properties as low-carbon steel, but has lower strength. It is tougher and more ductile and its internal structure is fibrous.

Cast iron is hard, brittle, less uniform in structure than wrought iron and steel, and also much weaker in tension.

Cast steel is far stronger than cast iron, especially in tension and resistance to impact.

Brass varies from a hard and brittle material to a ductile tenacious one, depending on the proportions of the various metals in the alloy.

Q What is the main difference between iron and steel?

A Chemically pure iron is an element, or simple substance, that cannot be divided up into anything but particles of the same substance. In practice, it is almost impossible to refine iron so that it is free from all foreign material, but a high-grade charcoal iron really contains few impurities. Steel is an alloy of iron and carbon. The carbon is in solution with the iron, increasing its strength and changing its physical properties to a marked extent.

Steel contains only a very small percentage of carbon, which is combined chemically with the iron. Cast iron, on the other hand, may contain quite a substantial amount of carbon, mostly in a "free" state, that is, simply *mixed* with the iron in the form of graphite.

Q What is the difference between low-carbon steel and high-carbon steel?

A Low-carbon steel contains a very small percentage of carbon, from 0.1 to 0.27 percent in soft malleable steel, such as is used for plates, bolts, and rivets, to as much as 0.6 percent in harder steels used in manufacturing rails, hard steel wire, etc.

High-carbon steel contains from 0.6 to about 1.5 percent. A 1 percent steel is suitable for such heavy-duty tools as chisels, punches, and rock drills. Turning tools, drills for metal, reamers, and dies are made from steel containing from 1 to 1.3 percent carbon.

Low-carbon steel is easily welded but cannot be hardened. High-carbon steel can be hardened but may not be so readily welded.

Q What is an alloy steel?

A An alloy steel contains some other metal or metals in addition to carbon and iron. Common alloying metals are chromium, nickel, manganese, silicon, vanadium, molybdenum, and tungsten. *Chromium* increases hardness and tenacity without reducing toughness. It is used extensively in high-speed tools and rustless steels; *nickel* greatly increases tensile strength, ductility, elasticity, and corrosion resistance; *manganese* increases hardness and toughness; *silicon* increases hardness and magnetic permeability, making steel particularly suitable for electromagnets; *vanadium* increases resistance to fatigue; *molybdenum* increases hardness and offers greater resistance to heat changes; and *tungsten* hardens and increases heat resistance. It is the main alloying constituent in self-hardening tool steels.

HEAT TREATMENTS AND PROCESSES _____

Q How is carbon tool steel hardened and tempered?

A Carbon tool steel is hardened by heating it to about 1450°F, then cooling it rapidly by plunging it into water or oil. The steel is not quite so hard and brittle when cooled in oil as when cooled in water, but in either case it is usually too hard and must be *tempered*—that is, some of the hardness drawn out. If the hardened steel's surface is polished and the steel reheated, colors appear, ranging from light straw to deep blue, as temperature increases. The darker the color, the higher the temperature and the softer the temper. If the tool is heated until the blue color appears, practically all hardness has been drawn from the steel. The tempering process can be stopped at any desired point by quenching the steel in oil or water.

Q List some common carbon-steel tools and their tempering colors.

A In the following tools (when made from carbon steel) the temper ranges from hardest to softest in the order given:
    LIGHT STRAW: Scrapers, files, lathe and planer tools for hard materials.
    STRAW: Milling cutters, lathe tools, taps, dies, reamers, drills.
    BROWN: Cold chisels for hard materials, woodworking tools.
    BROWNISH PURPLE: Cold chisels for soft materials, woodworking tools.
    DARK BLUE: Springs, wood saws.

Q What is meant by *annealing*?

A *Annealing* is the process of softening metal and relieving internal strains by heating to a temperature between the recrystallization temperature and the lower critical temperature. Rapid heating and short holding time prevent excessive grain growth. It is customary to cool the parts in air. If surface protection is desired, controlled cooling temperatures are used.

Q What is *case hardening*?

A This is a method of hardening the surface of those irons and steels that cannot be hardened by simply heating and then cooling rapidly. One method is to heat the metal to a cherry red, coat the surface with potassium cyanide or potassium ferrocyanide, and then chill suddenly by dipping in water. The cyanide carbonizes the surface of the metal, really converting it into a steel that can be hardened by heating and sudden cooling.

Q What is *soldering?*

A This is a process of joining two pieces of metal by covering their surfaces with a molten alloy of lead and tin, called *solder*. Parts to be joined are first cleaned to the bright metal, then coated with flux and "tinned" by spreading molten solder thinly over the surface with a soldering copper, blow torch, or other source of heat. More flux is applied and surfaces to be joined are clamped together and heated while additional solder is flowed into the joint. Parts are allowed to cool before pressure is released.

The flux helps the solder to flow and to form a firm bond with the metal being soldered. Common fluxes include rosin for tinned steel, rosin and tallow for lead, borax for iron, sal ammoniac for copper and brass. Chloride of zinc is a good general-purpose flux. In addition, several good prepared paste and liquid fluxes are on the market.

Q What is *brazing?*

A *Brazing* is somewhat similar to soldering, except that the strength and melting point of the joining material are much higher. While commonly called *spelter* by practical brazers, the material for high-temperature brazing is actually granulated brass with a melting point ranging from 1450 to 1800°F, according to its composition. So-called "low-temperature" brazing is done with silver alloys ranging in melting point from 1175 to 1600°F. Usual heat source is an oxyacetylene torch. Borax or other flux must be used. A brazed joint is very much stronger than a soldered joint.

Q What is *forge welding?*

A This is the joining of two pieces of metal by heating to their plastic point, then placing in contact, and hammering or pressing them together before they cool. When welding heat is reached, the metal is sprinkled with a flux to remove oxide from the molten surfaces and thus ensure a clean joint. A mixture of sand and borax or sal ammoniac is commonly used for this purpose. Strength of the weld depends on the cleanness of surfaces to be joined and the extent to which they really fuse. If temperature is not up to correct welding heat, metals will not fuse. Heated above welding heat, they will burn. A good weld is practically as strong as the solid metal.

Q Using sketches, explain what is meant by a *lap weld* and a *butt weld* (both forge welds).

A In *lap welding*, ends to be welded are first heated and "scarfed," as in Fig. 5-1a. They are then heated to welding temperature and welded by hammering, with ends overlapped as shown. In *butt welding*, ends are heated to welding temperature, joined by simply butting them together, as in Fig. 5-1b, and finished by hammering.

Q What is *fusion welding?*

A Any process that joins metals while in a molten state without hammering or other pressure is called *fusion welding*. Oxyacetylene and arc welding are examples of fusion welding.

Q What is *thermit welding?*

A Chemical action obtained by igniting a mixture of aluminum powder and iron oxide, called *thermit,* produces molten steel at a temperature of around 5000°F. This molten steel is run into a mold surrounding the parts to be welded, melting the parts, and solidifying with them as the steel cools. It is essentially a casting process.

Q Explain *oxyacetylene welding.*

A In the *oxyacetylene process* of fusion welding, the joint between parts to be welded is thoroughly cleaned and in many cases cut out in V shape. This vee is filled in by melting a rod of similar metal into the joint and, at the same time, fusing the edges of the metal to be joined with the metal of the filler rod. If fusing is properly done, the joint is practically as strong as the original metal. Heat required to melt the metal is applied by means of a blowpipe burning a mixture of oxygen and acetylene gas that produces an intensely hot flame. The flame must be kept moving with a circular or "weaving" motion to spread the heat and avoid burning the metal. Figure 5-2 shows how an oxyacetylene weld is made.

Q What are the principal methods of *electric welding?*

A The three common processes are:

RESISTANCE WELDING: Parts to be welded are raised to fusing tempera-

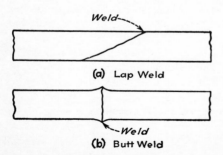

(a) Lap Weld

(b) Butt Weld

FIG. 5-1 Forge welding, lap and butt joints.

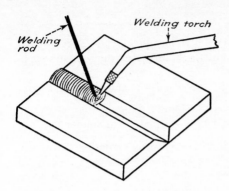

FIG. 5-2 Oxyactylene welding method.

ture by the passage of a heavy electric current while they are in contact with each other. Spot welding of light metal parts is one form of resistance welding.

CARBON-ARC WELDING: Parts to be welded are connected to one leg of an electric circuit; the other leg is connected to a carbon rod held in a hand receptacle. When the carbon rod touches the metal surfaces to be welded, and is then moved back a fraction of an inch, an arc is formed. Heat of the arc melts the metal, and a filler rod of similar metal held in the arc fills up the joint.

METALLIC-ARC WELDING: A wire electrode is used instead of the carbon rod, and the arc is formed between the rod and the metal to be welded. The electrode itself melts and fills the joint. Electrodes are bare wires or, more commonly, metal rods coated with some flux. Figure 5-3 shows this method of electric welding on a prepared V joint. The electrode is constantly moved in a weaving pattern during the welding process. The pattern varies with type of joint.

Q Is anyone allowed to make repairs on a boiler?
A *No.* Only qualified welders, and such repairs must be approved by an authorized inspector.

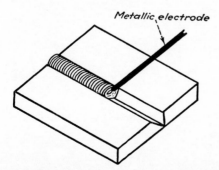

FIG. 5-3 Electric welding method.

FIGURING HEATING SURFACE

Since most of the parts that make up boiler-heating surfaces are circular or cylindrical shells, every engineer should know the easiest ways to figure these areas.

For practical engineering purposes, the most convenient formulas are:

Circumference of circle = 3.1416 × diamater
Area of circle = 0.7854 × square of diameter

Note that the second formula is just another way of stating that if you draw a square around a circle, the area of the circle is 78.54 percent of the area of the square. These rules are illustrated in Figs. 5-4 and 5-5.

It is not always necessary to use all known "places" or digits of a factor, piece of data, or result of a computation. How many to use depends on the accuracy required. For most boiler computations, it is close enough to use the factors 3.14 and 0.785, respectively. Using 3.14 instead of 3.1416 involves an error of only 0.05 percent of 1 sq ft out of 2,000 in boiler heating surface. Any attempt to figure closer than this is generally a waste of time because, first, 1 sq ft in 2,000 is not important, and second, errors in measurement throw the result off more than 0.05 percent in any case. The same applies to use of 0.785 instead of 0.7854.

FIGURING SURFACE AREA

The area of a flat rectangular surface is merely the product of its two dimensions. To figure the surface area of a cylinder, such as a tube or drum, imagine the surface unrolled to a flat, rectangular shape. Length doesn't change; width is the circumference of the original circle; area is the product, as shown in Fig. 5-5.

Here are some important points to remember in figuring heating surface:

1. If dimensions are in feet, area comes out in square feet.
2. If dimensions are in inches, area comes out in square inches. Divide square inches by 144 to get square feet.

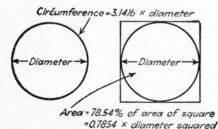

FIG. 5-4 Area of circle and square.

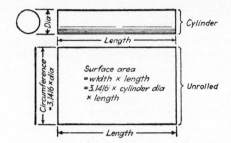

FIG. 5-5 Surface area of cylinder.

3. When many tubes are the same length, figure the surface of one tube in square feet. Then multiply this area by the number of such tubes.

4. Boiler-heating surface is always measured on the side exposed to the fire. Thus, the heating surface of *water tubes* is based on *outside tube diameter;* that of *fire tubes,* on *inside diameter.*

5. When tubes enter a sheet or drum, deduct area of the tube holes from the gross area of sheet or drum to get net heating surface.

Q How many square feet of heating surface are there in a boiler furnace flue 3 ft in diameter and 12 ft long?
A Circumference of flue = 3.14 × 3 = 9.42 ft. Heating surface = area of flue = circumference × length = 9.42 × 12 = 113.04, say 113 sq ft.

Q How many square feet of heating surface are there in an 18-ft-long boiler fire tube of 3¼ in. internal diameter?
A Heating surface of a fire tube must be based on *internal* circumference, which here is 3.25 × 3.14 = 10.205 in. This is 10.205 ÷ 12 = 0.8504 ft. Then, flue surface = 0.8504 × 18 = 15.307 sq ft, say, 15.31 sq ft.

COMMENTS ON DECIMAL PLACES

All through the last computation there is the question of how many "places" to carry the various intermediate answers. Many schools still teach that arithmetical problems should be carried a certain number of *decimal places,* but this is not sound practice and often gives foolish results — sometimes needlessly precise, sometimes too rough.

Examples make this point clear. The correct thickness of a certain piece of paper is, say, 0.0055 in. Any one taught to carry all numbers to three places of *decimals* would write down 0.005 in., which is in error by 10 percent. Asked to express a 1,000 in. length in feet, the same person would write 1,000 ÷ 12 = 83.333 ft, which is correct to better than 0.005 percent. It may well happen that this degree of precision has no practical meaning. The 83.333 is more than 2,000 times as precise as the 0.005, yet both have three decimal places. All of this proves that a given number of decimal places doesn't mean a given degree of precision.

What is important is not the number of *decimal places*, but the number of *significant figures*, so called. To determine the number of significant figures, start counting with the first digit *that is not a zero* and count all digits from there on, including zeros.

Thus 4163., 0.4163, and 0.0004163 are all written to four significant figures and all represent the same order of precision in percentages.

Similarly, 3 and 0.00000003 are both written to one significant figure only, and both represent a very *low* order of percentage precision, even though the second number is written to eight decimal places.

Such examples show that the engineer-reader who has been in the habit of counting decimal places should shift over to significant figures to keep out of trouble in computations.

In the problems worked out in the preceding heating surface examples, we used the value 3.14, a constant carried to three significant figures. That choice sets the limit of possible precision for the whole computation at about 0.05 percent. It is a good rule of thumb that the various steps in a computation may carry one or two more significant figures than the data or factors used. Anything beyond that is meaningless. Note that this rule was followed in working out the problems. Most of the steps were carried to four significant figures. The cases where the answer was carried to five figures were justified by the fact that the first digit was a low number. That often warrants using two more significant figures instead of one more. Thus 0.8504 was carried to four figures, but 10.205 to five figures.

FIRE-TUBE BOILER PROBLEM

We follow the above general rule in working out the following problem:

Q An hrt boiler is 5 ft in diameter and 16 ft long. It contains 60 tubes of 3 in. outside diameter and 2.732 in. inside diameter. Find the boiler heating surface. Take the inner surface of tubes, half of the shell surface, and two-thirds of the tube plate area, less the area of the tube holes.
A Circumference of shell = $5 \times 3.14 = 15.700$ ft

Half circumference of shell = $15.700 \times \frac{1}{2} = 7.850$ ft

Area of half of shell = $7.850 \times 16 = 125.60$ sq ft

Inner circumference of fire tube = $2.732 \times 3.14 = 8.578$ in.,

or $8.578 \div 12 = 0.7148$ ft

Surface of one tube = $0.7148 \times 16 = 11.437$ sq ft

Surface of 60 tubes = $60 \times 11.437 = 686.2$ sq ft

Total area of one tube sheet = 0.785 × 5 × 5 = 19.625 sq ft
Two-thirds area one tube sheet = $^2/_3$ × 19.625 = 13.083 sq ft
Area of one tube hole = 0.785 × 3 × 3 = 7.065 sq in.
Area of 60 tube holes = 60 × 7.065 = 423.9 sq in.,

or 423.9 ÷ 144 = 2.94 sq ft
Net heating surface of one tube sheet = 13.08 − 2.94 = 10.14 sq ft
Net surface of two tube sheets = 10.14 × 2 = 20.28 sq ft
Total net area = 125.60 + 686.2 + 20.28 = 832.08 sq ft,

say 832 sq ft. *Ans.*

SUGGESTED READING

REPRINT SOLD BY *Power* MAGAZINE,
    Corrosion, 36 pp.

BOOKS
    Elonka, Stephen M., and Anthony L. Kohan: "Standard Boiler Operators'
    Questions and Answers," McGraw-Hill Book Company, New York, 1969.
    Elonka, Stephen M.: "Standard Plant Operators' Manual," 2d ed., McGraw-Hill
    Book Company, New York, 1975.

6

BOILER CONSTRUCTION DETAILS

Although boilers today are welded, many older riveted boilers are still in service. Also, many examiners ask questions about riveted seams. Thus this chapter deals with the mechanics of materials in boilers in use today, and some details of boiler construction that the operator must know.

STRENGTH OF MATERIALS

Q What is meant by *load, stress,* and *strain?*

A *Load* is an *external* force acting on a body. *Stress* is the *internal* resistance that the particles composing a body offer to the action of an external load. *Strain* is a change of length or shape caused by an external load. There are three kinds of simple stress and strain: (1) tensile stress and tensile strain; (2) compressive stress and compressive strain; (3) shearing stress and shearing strain. Tensile or compressive strain is measured by the fraction: change of length ÷ original length.

Q If an iron bar 6 ft long and 2 sq in. in cross section is subjected to a pull of 20,000 lb and is stretched 0.024 in. in length, what are the load, unit stress, and strain?

A Load = 20,000 lb

Unit stress = load ÷ area = $20,000 ÷ 2 = 10,000$ psi

$$\text{Strain} = \frac{\text{increase in length}}{\text{original length}} = \frac{0.024}{72} = 0.00033 \text{ in. per in.}$$

Note that this last answer means there has been an extension in length of 0.00033 in. per in. of original length.

Q Explain what is meant by the following terms: *tensile unit stress, compressive unit stress,* and *shear unit stress.*

A *Stress* is the force acting within the material. *Unit stress* is the stress per square inch of cross section. It equals the external load divided by the area that carries the load. Thus, *tensile unit stress* in a rod under tension is equal to the total tension in pounds divided by the square inches of cross section. Likewise, *compressive unit stress* in a short rod under compression is the total compressive force in pounds divided by the cross-sectional area in square inches.

Shear unit stress is the total shearing force divided by the area resisting that shear. *All unit stresses,* whether tensile, compressive, or shear, are measured in pounds per square inch (psi). In Fig. 6-1, all four sketches, the unit stress is the total force divided by the area subjected to tension, compression, or shear. Although actual failure is pictured in each case, this merely suggests the type of failure that would occur if the type of stress listed were indefinitely increased. In Fig. 6-1c and d, note the difference between single and double shear. In the latter case, the area carrying the load is double the cross-sectional area of the rivet; therefore each rivet does double duty.

Q What is meant by the *elastic limit* of a material?

A If a tensile or pulling load is applied to a bar of metal, the bar stretches more and more as the load is increased, and for a while in exact proportion to the load. Finally, a load is reached at which the bar starts to

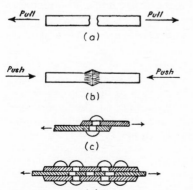

FIG. 6-1 Unit stresses. (*a*) Tensile; (*b*) compressive; (*c*) and (*d*) shear.

stretch faster for a given increase in load. The unit stress 'at this point is called the *elastic limit* of the material. If a bar is loaded to less than the elastic limit, it returns to its original length after the load is removed. If it is stretched beyond its elastic limit, it retains a permanent "set" after the load is removed.

Q What is the ultimate strength of a material?
A It is the total load (in tension, compression, or shear, as the case may be) required to pull the piece apart, to crush it, or to shear it, divided by the square inches of area resisting such failure. Since the cross section changes before a piece fails in tension or compression, the material's ultimate strength is customarily figured as the breaking load divided by the *original* cross section.

Q Is there any difference between the strength of a piece and the strength of the material of which it is made?
A If a certain steel has a tensile strength of 60,000 psi, a particular bar of 4 sq in. cross section has a strength of 240,000 lb. The same applies to elastic limit and to actual loads. In each case we may consider the *total* loading for an *object*, such as a bolt or a rivet, but only the unit stress (loading per sq in.) in the case of a *material*.

Q What is meant by the *safe working strength* of a material?
A This is the maximum unit stress deemed safe for a material to carry under ordinary working conditions. To provide a sufficient margin of safety, it is always far less than the elastic limit, which in turn is much less than the ultimate strength.

Q What is the *factor of safety*?
A It is the ultimate strength of the *material* divided by the *actual unit stress,* or it is the ultimate strength of the *piece* divided by the actual total load (tension, compression, or shear) on the piece. Thus, it is merely a number showing how strong the material or object is in terms of the imposed load use. If the factor of safety is 5, the load would have to be multiplied 5 times to break the piece.

Factor of safety must allow for uncertainties in the uniformity of the material, for corrosion and wear, for bumps and jerks, for fatigue, for overloading, and for imperfections in the method of figuring.

Thus, in practice, for each kind of work and material, the factors used are those which experience shows to be reasonably safe without undue waste of material from making parts overly strong.

Q What working steam pressure, in pounds per square inch, is allowed on a steam boiler whose bursting pressure is calculated to be 1,260 psi if the safety factor is 5.6?

A Note that the stress in the metal is unknown as far as this problem is concerned. Only the bursting steam pressure is given. However, stress in the metal is proportional to the steam pressure, so you merely divide 1,260 by 5.6 to get 225 psi, the safe working steam pressure.

Q A steel bolt carries a load that induces a unit stress of 6,000 psi. What is the ultimate strength of the bolt material if a safety factor of 8 is used?
A Ultimate strength = working load × factor of safety
$$= 6,000 \times 8 = 48,000 \text{ psi.}$$

Q If steel bars with an ultimate tensile strength of 56,000 psi are used to make boiler stays and the maximum permissible stress is 8,000 psi, what factor of safety is being used?
A Factor of safety
$$= \text{ultimate tensile strength} \div \text{maximum permissible stress}$$
$$= 56,000 \div 8,000 = 7.$$

Q What is the ultimate strength of a round steel bar, 1½ in. in diameter, if the steel's breaking strength is 60,000 psi?
A Area of bar = 0.785 × 1.5 × 1.5 = 1.77 sq in.
Ultimate strength = 60,000 × 1.77 = 106,200 lb.

Q What is the stress in six 1-in. bolts that support a load of 10 tons? Diameter at bottom of threads is 0.84 in.
A Total area of bolts = 6 × 0.84 × 0.84 × 0.785 = 3.32 sq in.
Stress = load ÷ area = (10 × 2,000) ÷ 3.32 = 6,024 psi.

Q What tests are applied to material that is used to manufacture a steam boiler, and what are the purposes of these tests?
A TENSION TEST: To find ultimate tensile strength of a material and amount of elongation when subjected to varying degrees of tensile stress.

CRUSH TEST: Applied to boiler tubes to see whether they can stand crushing longitudinally without cracking.

BEND TEST: To find whether metal stands bending without breaking or cracking.

FLATTENING TEST: Ability of part, when it is under test, to stand flattening without cracking.

TRANSVERSE TEST: Resistance of metal to bending.

HOMOGENEITY TEST: To find whether the internal structure of metal is uniform by examining the broken edges of a fractured specimen.

FRACTURE TEST: Same purpose as homogeneity test.

NICK BEND TEST: Same purpose as homogeneity test.

ETCH TEST: Examining structure of material by polishing the surface of the test specimen and etching the polished surface with acid.

HARDNESS TEST: For surface hardness.

HYDROSTATIC TEST: Water-pressure test applied to pipes, tubes, and castings to detect weak parts in structure; also applied to finished boilers and other pressure vessels to find weaknesses in material or defects in workmanship.

X-RAY TEST: To examine castings and welds for possible flaws.

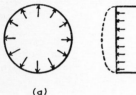
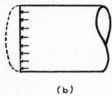

(a)   **(b)**

FIG. 6-2 Effect of pressure on cylindrical (*a*) and flat (*b*) surfaces.

Q Why are the principal parts of boilers subjected to internal pressure usually cylindrical in form?

A In a cylindrical vessel pressure is exerted equally in every direction on the cylindrical surface, and thus no part tends to become distorted through excessive pressure. See Fig. 6-2*a*. In Fig. 6-2*b*, pressure is shown acting on the flat head of a cylindrical shell. Edges of the flat plate are supported by the shell plate, but the center of the plate is not supported in any way and tends to bulge outward in an elliptical form, as the dotted line shows. Elliptical-shaped heads, such as those in many types of water-tube boiler drums, do not require support because they are already in the shape to which a flat head would be distorted by excessive pressure. On the other hand, flat plates *must* be stayed.

CONSTRUCTION

Q How are boiler plates joined?

A By welded or riveted joints.

Q What forms of rivet heads are allowed in boiler construction?

A Various acceptable forms of rivet heads are shown in the ASME Code for Power Boilers, and also in Fig. 6-3.

Q What forms of riveted joints were used in boiler construction?

A Lap joints and butt joints. In the lap joint, edges of plates overlap. In the butt joint, plate edges meet or butt together so that cover straps are used.

Q Describe and sketch a single-riveted lap joint.

A In this joint, plate edges are lapped over and secured by one row of rivets. Figure 6-4*a* shows a side view and end-sectional view of a single-riveted lap joint.

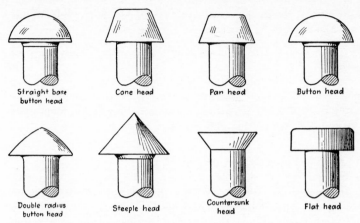

FIG. 6-3 Rivet heads are formed in these eight standard shapes.

Q Describe and sketch a double-riveted lap joint.
A In this joint, plate edges are lapped and secured by two rows of rivets, Fig. 6-4b.

Q Describe and sketch a double-riveted butt joint with double straps.
A In this joint, plate edges are butted together and cover straps are

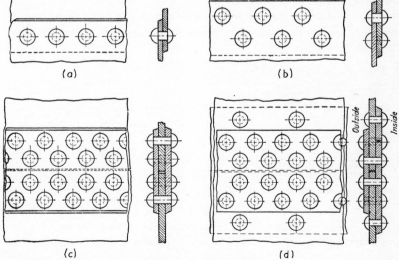

FIG. 6-4 Types of riveted joints used in older boilers.

placed inside and outside. Figure 6-4c shows a double-riveted double-strap joint with straps of equal width.

Q Describe and sketch a triple-riveted butt joint with double straps of unequal width.
A In this joint, Fig. 6-4d, plate edges butt together and cover straps are placed inside and outside, but the outside strap is narrower than the one inside. Outer rows of rivets pass through the inner strap and shell plate but not through the outer strap, and alternate rivets are omitted in these outer rows.

Q Why is a butt joint preferable to a lap joint?
A In the butt joint, shell plates form a true circle, and there is no tendency for the pressure to distort the plates in any way. In the lap joint, plates do not form a true circle at the lap, and internal pressure tends to pull them to the position shown in Fig. 6-5a. In time, this bending action may cause "grooving" or cracking of the plate along the calking edge and thus create a dangerous condition. Properly designed butt joints also have much higher efficiencies than lap joints with the same number of effective rows of rivets, because in the butt joint all or most of the rivets pass through three plate thicknesses and so are in *double shear,* while the rivets in lap joints are all in *single shear.* According to the ASME code, rivets in double shear have twice the strength of rivets in single shear.

Q What preparation of plates and butt straps is necessary to ensure good riveted joints?
A Plates and straps must be rolled or formed to the proper curvature by pressure only, and calking edges planed or milled to an angle not sharper than 70° with the plane of the plate. Plates and straps should fit closely

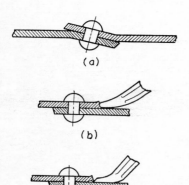

(a)

(b)

(c)

FIG. 6-5 (a) Distortion of single lap joint; (b) correct calking; (c) incorrect calking.

before riveting so as to avoid undue stress on the rivets or excessive calking to make a joint tight.

Q Are rivet holes punched or drilled?

A A drilled hole is best because it is parallel, exact in size, and the metal around it is not weakened to any great extent by the drilling process. The ASME code permits punching holes at least $\frac{1}{8}$ in. less than full diameter in plates not over $\frac{5}{16}$ in. thick, and at least $\frac{1}{4}$ in. less than full diameter in plates over $\frac{5}{16}$ in. and up to $\frac{5}{8}$ in. Drilling or reaming to full size removes metal around the hole that may have been crushed by the action of the punch.

Final drilling or reaming of rivet holes is done with the joint assembled and held in place by tack bolts. The parts are then separated and all burrs around holes, chips, and cuttings removed before the joint is reassembled and the rivets driven. Tack bolts and barrel pins keep the holes in proper line while the riveting is being done.

Q What is calking and why is it necessary?

A Calking refers to the upsetting or burring up of the edge of the plate or strap after riveting so as to make the edge press down tightly on the plate beneath and thus form a water-and steam-tight joint. This is necessary because inequalities in the plate surface make it practically impossible for riveting alone to make a joint absolutely tight. In calking, use a blunt-nosed tool and take great care not to damage or nick the bottom plate. A blunt or round-nosed tool thickens the plate edge, as in Fig. 6-5*b*, but a sharp tool may actually force the plates apart, as in Fig. 6-5*c*, and so do more harm than good. After calking and applying the hydrostatic test, certain seams may be seal-welded as a further precaution against leakage.

Q Which is stronger and safer, riveted or welded seams for pressure vessels?

A Experience and observation of burst tanks indicate that riveted seams in tanks should be avoided. Usually the joint or seam is the weakest part of the tank, whether a longitudinal or girth seam or a riveted head attached to the body. Bursting rivet heads and bodies cause a "shrapnel" effect, which is so forceful that the pieces can easily penetrate concrete walls and even pass through steel plating.

A welded tank conforming to ASME specifications (or to certain specifications of other societies) is generally considered safe. This also applies to boiler drums, although such riveted drums are rarely made today. In view of the damage that can occur just from flying rivet heads, it is best to always use welded drums that have been stress-relieved and built according to the applicable code.

Tanks (or flasks) holding compressed air or gas or liquid air or gas are probably the *most* dangerous should there be a fire or excess heat tending to increase the internal pressure. This applies especially to tanks that are only partially filled with volatile liquids; in such a case, the liquids more readily flash into gas than if the tank were completely full.

NONDESTRUCTIVE TESTING

Q What is a nondestructive test?

A As opposed to bending, pulling or twisting a piece of material until it breaks, nondestructive testing checks the soundness of the material without affecting it physically or chemically. Today these tests include radiography (x-rays), dye checks, and ultrasonic and hydrostatic testing. In power boiler testing, in addition to visual examination of the weld, x-rays and gamma rays are used and required. See the suggested reading at end of this chapter for more detailed information.

SUGGESTED READING

REPRINTS SOLD BY *Power* MAGAZINE,
  Power Handbook, 64 pp.
  Nondestructive Testing, 24 pp.

BOOKS
  "ASME Boiler and Pressure Vessel Code," Sections 1 and 9, American Society of
    Mechanical Engineers, New York, 1974.
  Elonka, Stephen M., and Anthony L. Kohan: "Standard Boiler Operators' Ques-
    tions and Answers," McGraw-Hill Book Company, New York, 1969.

7

STAYS, JOINTS, AND BOILER CALCULATIONS

So long as boilers contain flat surfaces, stays to keep them from bulging and bursting will be used. Knowledge of the types of stays and their construction, methods of inspecting the condition of stays when a boiler is opened, and calculations related to stays and stayed areas are all vital to the stationary engineer's work.

Whether joints are riveted as in older boilers, or welded as in modern units, the forces working to part them must be calculated before repairs are made. Such calculations are required in many license examinations. Completing the calculations in this chapter will give the operating engineer a better understanding of (and respect for) heated surfaces under pressure encountered in daily work—and, we hope, result in safer boiler operation.

STAYS _____

Q Why are stays necessary in boiler construction?

A Flat surfaces exposed to pressure tend to bulge outward; hence they must be supported by stays. Cylindrical or spherical surfaces do not tend to change their shape under pressure and so do not require staying.

Q Sketch a through stay and explain where it is used.

A Through stays are used to support widely spaced parallel flat surfaces, such as front and rear heads of return-tubular and dry-back boilers.

They are long round bars, threaded on the ends with a fine thread, and fitted with nuts and washers to secure the ends of the stay to plates and make a steamtight and watertight joint. When the outer nut would be exposed to excessive heat (as, for example, on the rear head of a return-tubular boiler), an eye is forged on the end of the stay; it is fastened to the rear head by a pin and angle irons. To avoid exposing a double thickness of metal to the furnace heat at this point, the angle iron may be separated a few inches from the head by distance pieces riveted to the angle iron and to the head. Both methods are shown in Fig. 7-1.

Q   Sketch and describe a diagonal stay and explain where it is used.
A   It is used for the same purpose as the through stay — that is, to stay the flat portions of cylindrical-shell heads that are not supported by the tubes. It is not so direct as the through stay, and it throws stress on the shell plates as well, but it leaves more room above the tubes for inspection, repair, and cleaning. A common form of diagonal stay is shown in Fig. 7-2.

Q   Sketch and describe a gusset stay.
A   It is a form of diagonal stay in which a plate is used instead of a bar. As illustrated in Fig. 7-3, it consists of a heavy plate fastened by rivets and angle bars to the head and shell. It is more rigid than the diagonal stay, takes up more room, and interferes to a greater extent with water circulation. Gusset stays are used very little in modern boiler construction.

Q   Sketch and describe a girder stay.
A   The girder stay is still used to support tops of combustion chambers in boilers of the Scotch marine type. It consists of a cast steel or built up girder with its ends resting on the side or end sheets of the firebox or combustion chamber, and supporting the flat crown sheet or top sheet of the combustion chamber by means of bolts, as shown in Fig. 7-4.

Q   How are curved crown sheets usually supported in older boilers?
A   They are supported by long threaded rods called *radial stays,* Fig. 7-5. These rods are screwed through both crown sheet and wrapper sheet and the ends are riveted over.

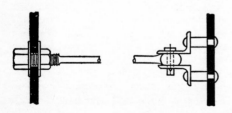

FIG. 7-1 Methods of attaching through stays to boiler plate.

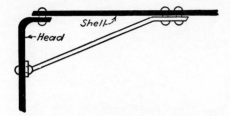

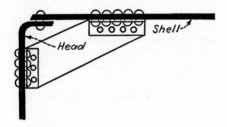

FIG. 7-2 Diagonal stay application.

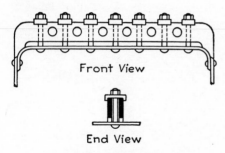

FIG. 7-3 Gusset stay is strong.

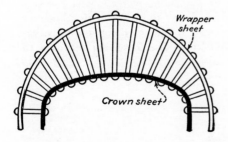

FIG. 7-4 Girder stay for crown sheet of wet-back Scotch marine-type boiler.

FIG. 7-5 Radial stays for crown sheet of firebox boiler.

Q Sketch and describe some forms of stay bolts.

A They are short stay bars to support flat surfaces that are only a short distance apart, such as inner and outer sheets of the water legs in a locomotive boiler. A plain stay bolt (*a*), Fig. 7-6, is simply a piece of round

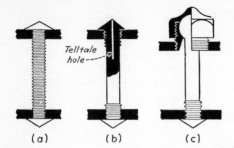

FIG. 7-6 Types of stay bolts for fire-tube boilers.

iron or steel bar, threaded its entire length with a fine thread (12 threads per inch) screwed through both sheets and riveted over on each end. The stay bolt at (*b*) has the threads turned off on the center part, or sometimes bolt ends are upset to give enough extra diameter to form a thread and leave the center part smooth. This type is a little more flexible than (*a*) and should be more durable because the bottom of a sharp V thread is a very likely place for a crack to start. It is also claimed to be less liable to damage by corrosion than the all-threaded stay bolt. A still more flexible form, with ball and socket joint on one end, is shown at (*c*).

Small telltale holes are sometimes drilled in stay-bolt ends to a depth greater than the thickness of head and plate, so that broken stay bolts can be detected by the escape of steam and water. In the case of solid stay bolts, broken bolts can be detected by tapping the heads with a hammer and noticing the sound, but this requires some practice.

Q Sketch some methods of connecting inner and outer plates at the bottom of water legs and openings around fire doors or burner openings in some types of boilers.

A Figure 7-7 shows four methods of connecting plates in such cases: (*a*) an ogee connection; (*b*) a solid ring between the plates with connecting rivets passing through ring and plates; (*c*) plates flanged inward and welded; (*d*) plates flanged inward but with one flange lapped over the other and welded as shown.

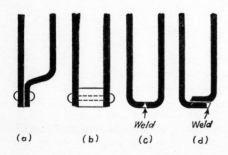

FIG. 7-7 Methods of connecting plates in leg-type fire-tube boilers.

RIVETED JOINTS

Q What is meant by the *efficiency* of a riveted joint?
A This is the ratio of the strength of a unit section of the joint to the same unit length of solid plate. Unit length usually taken is the pitch of the rivets (distance from center to center) in the row having the greatest pitch.

Q What are the various ways in which a riveted joint may fail?
A It may fail by (1) tearing the plate along the center line of the rivets, (2) shearing the rivets, (3) crushing plates and rivets.

Q What symbols are commonly used in riveted-joint calculations?
A Following are the symbols commonly used in riveted-joint calculations and their meanings: P = pitch of rivets, in., in the row having the greatest pitch; d = diameter of rivet after driving, or diameter of rivet hole, in.; a = cross-sectional area of rivet after driving, sq in.; t = thickness of plates, in.; b = thickness of butt strap, in.; s = shearing strength of rivet in *single shear,* usually taken as 44,000 psi; S = shearing strength of rivet in *double shear,* usually taken as 88,000 psi; c = crushing strength of mild steel, usually taken as 95,000 psi; T_s = tensile strength of plate metal, usually taken as 55,000 psi; n = number of rivets in *single shear;* N = number of rivets in *double shear.*

Q What is meant by *single shear* and *double shear?*
A When two lapping plates are riveted, as in a lap joint or a single-strap butt joint, there is only one section of each rivet opposing the tendency to shear the rivets through at right angles to their length. In this case, rivets are said to be in *single shear.* When there are three plate thicknesses in the joint, as in a butt joint with inner and outer straps, two sections of each rivet oppose the shearing stress, so the rivets are said to be in *double shear.*

Q How would you find the efficiency of a single-riveted lap joint?
A Figure 7-8a shows this joint. There are four calculations to be made, using the symbols given previously (the calculations refer to a length of joint equal to P, the rivet pitch):

1. Strength of solid plate = $P \times t \times T_s$
2. Strength of plate between rivets = $(P - d) \times t \times T_s$
3. Shearing strength of *one* rivet in *single shear* (since there is only one rivet in one pitch in this case) = $s \times a$
4. Crushing strength of plate in front of *one* rivet = $d \times t \times c$

Whichever one of 2, 3, or 4 gives the lowest value is divided by No. 1 to give the joint efficiency.

Q What is the efficiency of a single-riveted lap joint having a rivet pitch of 3 in., a rivet diameter of 1¼ in., and a plate thickness of ½ in.?

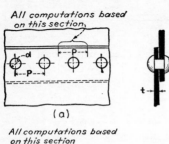

All computations based on this section

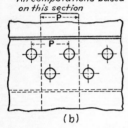

(a)

All computations based on this section

(b)

FIG. 7-8 P indicates pitch in single-riveted (a) and double-riveted (b) lap joints.

A Using substitutes in the formulas given in the preceding answer, the following are obtained (carrying each computation to nearest 100 lb):

1. $P \times t \times T_s = 3 \times 0.5 \times 55{,}000 = 82{,}500$ lb
2. $(P - d) \times t \times T_s = (3 - 1.25) \times 0.5 \times 55{,}000 = 48{,}100$ lb
3. $s \times a = 44{,}000 \times 1.25^2 \times 0.785 = 54{,}000$ lb
4. $d \times t \times c = 1.25 \times 0.5 \times 95{,}000 = 59{,}400$ lb.

Since (2) gives the lowest value, efficiency of joint = 48,100/82,500 = 0.583 or 58.3 percent. *Ans.*

This means that plate will fail by tearing the "ligament" between rivet holes and that the strength of this ligament is 58.3 percent of the strength of the undrilled plate.

Q How do you find the efficiency of a double-riveted lap joint?

A Here (Fig. 7-8b), calculations are similar to those for the single-riveted lap joint, except that the value of n is 2 instead of 1, because there are *two* rivets in one pitch in a double-riveted lap joint. Thus,

1. Strength of solid plate $= P \times t \times T_s$
2. Strength of plate between rivet holes $= (P - d) \times t \times T_s$
3. Shearing strength of *two* rivets in *single* shear $= 2 \times s \times a$
4. Crushing strength of plate in front of *two* rivets $= 2 \times d \times t \times c$

Q What is the efficiency of a double-riveted lap joint having a rivet pitch of 4 in., a rivet diameter of 1 in., and a plate thickness of ⅜ in.?

A Using formulas given in preceding answer and evaluating to nearest 100 lb

1. $P \times t \times T_s = 4 \times 0.375 \times 55{,}000 = 82{,}500$ lb

2. $(P - d) \times t \times T_s = (4 - 1) \times 0.375 \times 55,000 = 61,900$ lb
3. $2 \times s \times a = 2 \times 44,000 \times 0.785 = 69,100$ lb
4. $2 \times d \times t \times c = 2 \times 1 \times 0.375 \times 95,000 = 71,300$ lb.

Since (2) (strength of plate between rivet holes) is lowest, efficiency of joint $= 61,900/82,500 = 0.75$ or 75 percent. *Ans.*

Q How do you find efficiency of a triple-riveted butt joint with two unequal straps and alternate rivets omitted in outer rows?
A This (Fig. 7-9) is the equivalent of five rivets in one pitch, four being in double shear and one in single shear. Since there are a greater number of rows of rivets than in the previous riveted-joint examples, some extra calculations have to be made, as follows:

1. Strength of solid plate $= P \times t \times T_s$
2. Strength of plate between rivet holes in outer row $= (P - d) \times t \times T_s$
3. Shearing strength of *four* rivets in *double shear* and *one* rivet in *single shear* $= 4 \times S \times a + s \times a$
4. Strength of plate between rivet holes in second row, plus shearing strength of one rivet in single shear in outer row $= (P - 2d) \times t \times T_s + s \times a$
5. Strength of plate between rivet holes in second row, plus crushing strength of butt strap in front of one rivet in outer row $= (P - 2d) \times t \times T_s + d \times b \times c$
6. Crushing strength of plate in front of four rivets, plus crushing strength of butt strap in front of one rivet $= 4 \times d \times t \times c + d \times b \times c$
7. Crushing strength of plate in front of four rivets, plus shearing strength of one rivet in single shear $= 4 \times d \times t \times c + s \times a$

Q What is the efficiency of a triple-riveted butt joint with unequal straps

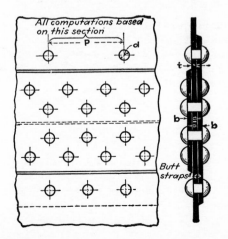

FIG. 7-9 P indicates pitch in triple-riveted butt joint.

and alternate rivets omitted in outer rows, if the outer pitch is 7 in., rivet diameter 1 in., thickness of plates $^{7}/_{16}$ in., and thickness of butt strap $^{3}/_{8}$ in.?

A 1. $P \times t \times T_s = 7 \times 0.4375 \times 55,000 = 168,400$ lb

2. $(P - d) \times t \times T_s = (7 - 1) \times 0.4375 \times 55,000 = 144,400$ lb

3. $4 \times S \times a + s \times a = 4 \times 88,000 \times 0.785 + 44,000 \times 0.785 = 310,900$ lb

4. $(P - 2d) \times t \times T_s + s \times a = (7 - 2) \times 0.4375 \times 55,000 + 1 \times 44,000 \times 0.785 = 154,900$ lb

5. $(P - 2d) \times t \times T_s + d \times b \times c = (7 - 2) \times 0.4375 \times 55,000 + 1 \times 0.375 \times 95,000 = 155,900$ lb

6. $4 \times d \times t \times c + d \times b \times c = 4 \times 1 \times 0.4375 \times 95,000 + 1 \times 0.375 \times 95,000 = 201,900$ lb

7. $4 \times d \times t \times c + s \times a = 4 \times 1 \times 0.4375 \times 95,000 + 1 \times 44,000 \times 0.785 = 200,800$ lb

Since calculation 2 gives the lowest result, efficiency of joint = 144,400/168,400 = 0.857 or 85.7 percent. *Ans.*

Q What will be the maximum allowable working pressure in psi on a cylindrical boiler shell if the inside diameter of the weakest course is 6 ft, plate thickness is $^{9}/_{16}$ in., efficiency of riveted joint is 85 percent, maximum allowable unit stress on steel is 55,000 psi, and maximum temperature to which plate will be subjected is 650°F?

A From the conditions given, the ASME Code for Power Boilers will allow a maximum unit working stress of 11,000 psi, and the maximum allowable working pressure is calculated from the code formula,

$$P = \frac{SEt}{R + 0.6t}$$

where P = maximum allowable working pressure, psi
S = maximum allowable unit working stress, psi
E = efficiency of longitudinal joint
t = minimum thickness of shell plates in weakest course, in.
R = inside radius of weakest course of shell, in.

Substituting the values given,

$$P = \frac{11,000 \times 0.85 \times 0.5625}{36 + 0.6 \times 0.5625} = 144.7, \text{ say 145 psi} \textit{Ans.}$$

Q In a fire-tube boiler, 72 in. in diameter and 18 ft long, the distance from the top of the tubes to the shell is 23 in. What is the area of the segment, Fig. 7-10, of the head above the tubes to be supported by staying?

A It is usually assumed that a distance of 2 in. above the top of the tubes is supported by the tubes and that a distance of 3 in. from the shell is supported by the shell. The tables of areas of segments given in the ASME Code for Power Boilers are based on these assumptions though the code also gives formulas for finding the distance d supported by the shell. These formulas are

(1) $d =$ outer radius of the flange, not exceeding 8 times the thickness of the head,

or

(2) $d = \dfrac{5T}{\sqrt{P}}$

where $T =$ thickness of the head, sixteenths of an inch
$\quad\quad P =$ maximum allowable working pressure, psi
   Assuming in this case that $T = 9$ and $P = 145$, then $d = 5 \times 9/\sqrt{145} = 3.8$ in., but for all practical purposes, 3 in. may be taken in most cases. The height of the segment to be stayed is $h = H - d - 2$, and the complete formula for finding the area is

$$\text{Area of segment} = \frac{4}{3}\,(H - d - 2) \times \sqrt{\frac{2(R - d)}{H - d - 2} - 0.608}$$

We can simplify this by substituting h for $H - d - 2$. Then

$$\text{Area} = \frac{4 \times 18^2}{3} \times \sqrt{\frac{2(36 - 3)}{18} - 0.608}$$

$$= 432 \times \sqrt{\frac{66}{18} - 0.608}$$

$$= 432 \times \sqrt{3.0587} = 432 \times 1.75$$

$$= 756 \text{ sq in.} \quad Ans.$$

Q Calculate the number and size of through stays required to support the segment in the previous question.
A As we cannot have two unknown quantities in our problem we must assume either the size of the stays or the number of stays to begin with. Let us assume that $1\frac{1}{2}$-in. stays are to be used, having an allowable unit stress of 8,500 psi, and that gage pressure is 145 psi.

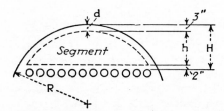

FIG. 7-10   Area of segment being calculated.

Allowable load on one stay = cross-sectional area × 8,500
$$= 1.5^2 \times 0.785 \times 8,500 = 15,013 \text{ lb}$$
Total load on segment, = area in sq in. × gage pressure
$$= 756 \times 145 = 109,620 \text{ lb}$$
$$\text{Number of stays} = \frac{109,620}{15,013} = 7.3, \text{ say 8 stays} \quad Ans.$$

Q Assuming that this segment is to be supported by diagonal stays instead of through stays, what would be their cross-sectional area if the stays are 40 in. long and the distance l in Fig. 7-11 is 35 in.?
A The ASME Code for Power Boilers formula for finding the area of a diagonal stay is

$$A = \frac{aL}{l}$$

where A = cross-sectional area of diagonal stay
    a = cross-sectional area of through stay
    L = length of diagonal stay
    l = length of line drawn at right angles to boiler head, or surface
        supported to center of palm of diagonal stay

Then, in this case,

$$\text{Area of diagonal stay} = \frac{1.767 \times 40}{35} = 2.02 \text{ sq in.}$$

and, if the stay is a round bar,

$$\text{Diameter} = \sqrt{\frac{2.02}{0.785}}$$

$$= 1.6 \text{ or } 1\tfrac{5}{8} \text{ in. to nearest eighth of an inch} \quad Ans.$$

Q Figure 7-12 shows a form of cylindrical flue for internally fired boilers known as an Adamson ring. If the diameter of the flue is 36 in. and the length of the section is 48 in. between ring centers with a plate thickness of $\tfrac{5}{8}$ in., what will be the allowable working pressure on this flue in psi?

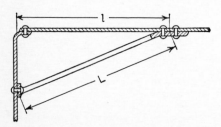

FIG. 7-11 Finding size of diagonal stay.

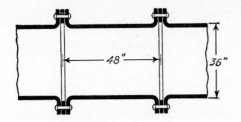

FIG. 7-12 Adamson ring-type combustion chamber.

A The allowable working pressure on an Adamson ring is found from the formula

$$P = \frac{57.6}{D} (18.75T - 1.03L)$$

where P = maximum allowable working pressure, psi
  D = outside diameter of the furnace, in.
  L = length of furnace section, in.
  T = thickness of plate, sixteenths of an inch
In this case,

$$P = \frac{57.6}{36} (18.75 \times 10 - 1.03 \times 48)$$

$$= 1.6(187.5 - 49.44) = 1.6 \times 138.06 = 220 \text{ psi} \quad Ans.$$

Q What is the maximum allowable working pressure on the concave side of an unstayed seamless head, dished to the form of a segment of a sphere of radius 42 in.? The plate is ⅝ in. thick and the maximum allowable unit working stress on the plate is 12,000 psi.
A The ASME Code for Power Boilers formula for plate thickness is

$$t = \frac{5PL}{6SE}$$

where t = thickness of plate, in.
  P = maximum allowable working pressure, psi
  L = radius to which head is dished, measured on concave side, in.
  S = maximum allowable working stress, psi
  E = efficiency of weakest joint forming head (efficiency of seamless heads is 100 percent)
Transposing the above formula for P and substituting the values given,

$$P = \frac{6tSE}{5L} = \frac{6 \times 0.625 \times 12,000 \times 1}{5 \times 42}$$

$$= 214 \text{ psi} \quad Ans.$$

Q What working pressure should be allowed on stayed flat plates, $3/8$ in. thick, with stay bolts pitched 5 in. horizontally and vertically?
A Working pressure is found from the formula in the ASME Code for Power Boilers,

$$Wp = \frac{CT^2}{p^2}$$

where C = a constant = 112 for screwed stays and plates not over $7/16$ in. thick
T = thickness of plates, sixteenths of an inch
p = pitch of stays, in.
Then,

$$Wp = \frac{112 \times 6^2}{5^2} = 161 \text{ psi} \quad Ans.$$

Q A flat plate, $3/8$ in. thick, is to carry 160 psi working pressure; what should be the pitch of the stay bolts?
A Using the same formula as in the previous question but transposing it for p,

$$p = \sqrt{\frac{CT^2}{Wp}} = \sqrt{\frac{112 \times 36}{160}} = 5 \text{ in.} \quad Ans.$$

Q A steam drum in a bent-tube boiler is 40 in. in diameter and made from $5/8$ in. carbon steel plate. A $1/2$-in. reinforcing plate is riveted over the ligament between the tubes, and the ligament has an efficiency of 45 percent. The longitudinal seam of the drum is a double-riveted butt joint with double straps and has an efficiency of 82 percent. The allowable unit working stress on the plate is 9,500 psi. What is the maximum allowable working pressure on this drum?
A Using the ASME Code for Power Boilers formula, Wp calculated on efficiency of the joint

$$= \frac{SEt}{R + 0.6t} = \frac{9,500 \times 0.82 \times 0.625}{20 + 0.6 \times 0.625}$$

$$= \frac{95 \times 82 \times 0.625}{20,375} = 239 \text{ psi}$$

Wp calculated on efficiency of the ligament

$$= \frac{9,500 \times 0.45 \times 1.125}{20 + 0.6 \times 1.125} = \frac{95 \times 45 \times 1.125}{20.675}$$

$$= 233 \text{ psi}$$

Taking the lower value, the allowable working pressure is 233 psi. *Ans.*

Q Calculate number and pitch of 1-in. stay bolts required to support the flat side sheet of a water leg measuring 30 by 42 in. Boiler pressure is 175 psi, and allowable unit stress on stay bolts is 7,500 lb per sq. in.

A Area of sheet = $30 \times 42 = 1,260$ sq in.

Total pressure on sheet = $1,260 \times 175 = 220,500$ lb

Assuming 1-in. stays are used, 12 threads per inch, U.S. standard,

$$\text{Diameter of stay at root of thread} = \text{outside diameter}$$
$$- (\text{pitch} \times 1.732 \times 0.75)$$
$$= 1 - (\frac{1}{12} \times 1.732 \times 0.75)$$
$$= 1 - 0.108 = 0.892 \text{ in.}$$
$$\text{Cross-sectional area of one stay} = 0.892^2 \times 0.785$$
$$= 0.625 \text{ sq in.}$$

$$\text{Total area of stays} = \frac{\text{total load}}{\text{allowable load per sq in.}}$$

$$= \frac{220,500}{7,500} = 29.4 \text{ sq in.}$$

$$\text{Number of stays} = \frac{29.4}{0.625} = 47,$$

say 48 to get even
spacing of six rows of eight
stays in each row   *Ans.*

$$\left.\begin{array}{l}\text{Vertical pitch} = {}^{30}\!/_6 = 5 \text{ in.}\\ \text{Horizontal pitch} = {}^{42}\!/_8 = 5\frac{1}{4} \text{ in.}\end{array}\right\} \quad Ans.$$

Q What weight of water will be contained in an hrt boiler, 18 ft long by 72 in. in diameter with seventy-two 4-in. tubes, when it is filled with water for a hydrostatic test?

A Volume of shell = $6^2 \times 0.785 \times 18 = 508.68$ cu ft

$$\text{Volume of tubes} = \frac{72 \times 4^2 \times 0.785 \times 18}{144} = 133.04 \text{ cu ft}$$

Volume of water = $508.68 - 113.04 = 395.64$ cu ft

Weight of water = $395.64 \times 62.4 = 24,688$ lb   *Ans.*

SUGGESTED READING

Reprint sold by *Power* magazine,
   Power Handbook, 64 pp.

Book
   Elonka, Stephen M., and Anthony L. Kohan: "Standard Boiler Operators' Questions and Answers," McGraw-Hill Book Company, New York, 1969.

8

SAFETY APPURTENANCES, COMBUSTION CONTROL

Among the more common preventable accidents are those involving overheating of a boiler. Most of these occurrences are caused by low water conditions. Of the accidents in a recent 1-year period, 89 percent of the accidents to low-pressure fire-tube heating boilers, 69 percent of accidents to high-pressure fire-tube power boilers, and 42 percent of accidents to all water-tube boilers involved overheating or burning of metal.

To avoid such accidents, boilers are equipped with safety appurtenances, protective controls, etc. Nevertheless, of the accidents listed above, the failure of operating or protective controls to function properly accounted for 52 percent of the accidents to low-pressure fire-tube heating boilers, 46 percent of the accidents to high-pressure fire-tube power boilers, and 16 percent of the accidents to water-tube boilers.

In this chapter, in addition to learning about safety appurtenances and controls, we explore *why* these accidents happen, and *how* to prevent them.

PRESSURE GAGES

Q What are safety appurtenances or boiler trim?
A They are valves, gages, and other connections or devices attached directly to the boiler and necessary for safe and efficient operation.

Q Name the most important fittings on a steam boiler and explain their uses briefly.

A Their names and uses are: *Safety valve* prevents dangerously high pressure and possible explosion by relieving boiler of excess steam. *Steam gage* indicates pressure of steam in boiler. *Water gage* shows water level in drum. *Gage* or *try cock* tests water level. *Fusible plug* prevents damage to boiler by giving warning of low water. *Stop valve* controls steam flow from boiler, cutting it on or off line as required. *Feed connection* includes check valve, stop valve, and piping needed to convey and control feedwater to boiler. *Blowoff connection* includes blowoff valves and pipe to reduce concentration of soluble impurities in boiler water, to blow out mud or sediment, and to drain boiler when necessary.

Q What types of gages indicate steam pressure?

A Two main types of pressure gages are *bourdon* and *diaphragm*. Figure 8-1*a* is a single-tube bourdon gage with dial removed to show the interior mechanism. The bent bourdon tube of oval cross section is closed at one end and connected at the other to boiler pressure. The closed end of the tube is attached by links and pins to a toothed quadrant, which in turn meshes with a small pinion on the central spindle. When pressure is applied to interior of oval tube, it tends to assume a circular cross section, but before the tube can do so it must straighten out. This tendency to straighten moves the free end, turning the spindle by the links and gearing, and causing the needle to move and register the pressure on a graduated dial.

The double-tube bourdon gage, Fig. 8-1*b*, is more rigid than the single-tube. It is more suitable for portable boilers.

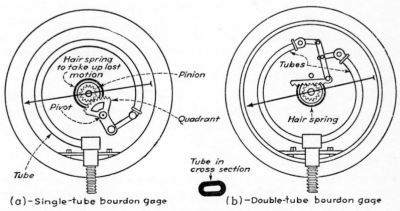

(a)-Single-tube bourdon gage   (b)-Double-tube bourdon gage

FIG. 8-1 Bourdon-tube steam-pressure gages may be single- or double-tube.

In the diaphragm gage a corrugated flexible metal diaphragm is clamped tightly between two flanges and connected to a toothed quadrant which meshes with a small pinion on the central pointer spindle.

Recording gages have a mechanism similar to that of tube and diaphragm gages described, but a revolving chart operated by clockwork takes the place of the stationary dial, and a pen which traces a record on the chart is attached to the end of the pointer.

Q What is the purpose of the steam gage, and how should it be connected to the boiler?

A The steam gage shows steam pressure in pounds per square inch. Connect it directly to the boiler at the highest point of the steam space or to the water column or water column-steam connection. The steam gage should be graduated to at least $1\frac{1}{2}$ times the set pressure of safety valve. It should be well-lighted and in plain view of the operator. A siphon or equivalent device must be placed between gage and boiler, and a tee or lever-handle cock provided near the gage to shut it off from the boiler when necessary. The tee or lever handle of this shutoff cock must be parallel to the pipe connections to the siphon when cock is open.

If the steam gage is directly connected to the boiler, no other valve may be placed on the line from boiler to gage unless length of pipe exceeds 10 ft, in which case a shutoff cock, or valve, arranged so it can be sealed open, may be placed near the boiler.

Q What is a siphon?

A This device prevents steam from entering the steam gage. Three forms are shown in Fig. 8-2: *Pipe* siphon (*a*) is simply a piece of pipe, bent as shown. *Ball* or *radiator* siphon (*b*) is a hollow-cast ball containing two small tubes attached to the inlet and outlet. The third form (*c*) is a U tube. In all three, water collects because of steam condensation, and the water transmits pressure to the gage, at the same time preventing fresh steam from entering the gage. As a siphon cannot drain itself, remove or drain it when the boiler is being shut down for a lengthy period and there is danger of freezing.

The ASME code requires that each boiler have a $\frac{1}{4}$-in. pipe-size connection to attach a test gage for checking the boiler steam gage while the boiler is in service. Connections for this purpose are shown in Fig. 8-2 between siphon and boiler steam gage.

Q How are steam gages tested?

A They are tested in three ways: (1) by comparison on the boiler with inspector's test gage; (2) by a screw plunger pump; (3) by a deadweight plunger pump.

In the first method, the test gage is attached to the boiler connection

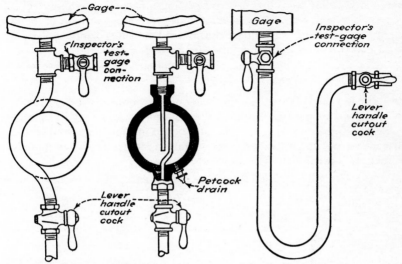

(a)-Pipe siphon (b)-Radiator siphon (c)-U-tube siphon

FIG. 8-2 Types of siphons used to protect steam pressure gages.

for this purpose and the two gages are compared as boiler pressure rises or falls.

In the second method, Fig. 8-3a, test gage and gage to be tested are attached to a T-shaped connection at the bottom of a cylinder, fitted with a

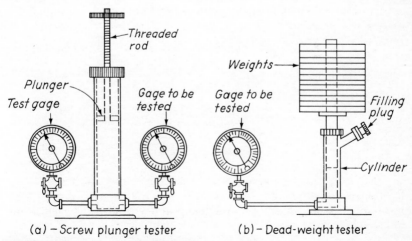

(a) – Screw plunger tester (b) – Dead-weight tester

FIG. 8-3 Testers used for pressure gages.

plunger attached to a threaded rod passing through the cylinder cover. To operate, unscrew top cover, fill cylinder with water or light oil, screw on top cover, and force plunger downward by turning handle on end of threaded rod. This puts equal pressure on both gages.

In the deadweight tester, Fig. 8-3*b*, the plunger, working in a cylinder, floats on oil and is loaded by weights.

Q Why is a siphon used between boiler and steam-pressure gage?
A The siphon prevents steam from entering the gage. If steam were allowed to pass into the gage tube of a bourdon gage, high temperature would expand the tube and cause a false reading. High temperature might also have a bad effect on metal.

Q What is meant by *gage pressure* and *absolute pressure*?
A *Gage pressure* is pressure *above* atmospheric pressure. *Absolute pressure* is gage pressure plus atmospheric pressure. Atmospheric pressure is 14.7 psi at sea level; it decreases as height above sea level increases. Gages on steam boilers indicate gage pressure.

WATER GAGES AND COLUMNS

Q What is the purpose of a water gage and how is it attached to a steam boiler?
A The water gage shows water level. In such boilers as the vertical-tubular types it is usual to attach water-gage glass fittings directly on the boiler head or shell, but in dry-back marine boilers the setting or smoke-box prevents direct attachment, so a water column is used. Water columns are also used on water-tube boilers where direct attachment to drums is not convenient.

Q Sketch the water gage connections on a firebox- or water-leg-type boiler. What is the position of the gage with reference to the highest point of the boiler heating surface, and how would you test this gage to prove that the passages leading to the glass are clear?
A Figure 8-4 shows the position of the gage on the boiler front. The visible bottom of the glass must be at least 3 in. above the highest point of the crown sheet.

To prove that the passages leading to the glass are clear, the following steps are taken:

1. Close top steam valve and open drain valve, thus allowing water to blow through the bottom passage and prove it to be clear.

2. Close bottom water valve and open top steam valve, allowing steam to blow through and prove steam passage to be clear.

3. Open bottom water valve and close drain valve. If all passages are

clear and unobstructed, the water will rise quickly to its correct level in the glass when the drain valve is closed.

This is the proper procedure in testing the water level, since simply opening the drain valve, without touching any of the other valves, does not prove that both top and bottom connections to the glass are clear.

Q What is a water column, and how should it be connected to a boiler? Sketch a water column and its connections.

A A water column is a hollow casting, or forging, connected by pipes at top and bottom to the boiler's steam and water spaces, Fig. 8-5. The steam-pipe connection to the top of the water column must not be lower than the top of the glass, and the water-pipe connection to the column must not be higher than the bottom of the glass. The minimum size of these connecting pipes must not be less than 1 in. Use plugged tees or crosses wherever practicable at right-angle turns, so that all parts of the piping may be easily examined and cleaned by removal of plugs.

Valves are not absolutely essential on steam and water connections to the water column, but, if used, they must be outside screw and yoke, lever-lifting gate valves, stopcocks with lever handles, or other valve types that offer a straight-way passage and show by the position of the operating mechanism whether they are open or closed. Lock these valves or cocks open or see that they are sealed open.

The water-gage glass with its steam, water, and drain valves is placed on the water column as shown, along with the required number of gage cocks. Damper regulators, feed-water regulators, steam gages, and other pieces of apparatus that do not require or permit escape of an appreciable amount of steam or water may be connected to the pipes leading from water column to boiler. Cast-iron water columns may be used for

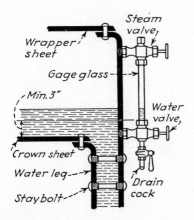

FIG. 8-4 Water gage on firebox- or water-leg-type boiler.

pressures not exceeding 250 psi, and malleable-iron columns for pressures not exceeding 350 psi. Above that, use steel columns.

Q How do you test the water column and water-gage glass to prove that all passages are clear, while the boiler is in operation?

A 1. Close top valve on column and top valve on glass. Open drain valve on glass. If water blows freely from drain, the water passages from boiler to column and from column to glass are clear.

2. Close bottom valves on column and glass, and open the top valves. If steam blows freely from drain valve at bottom of glass, steam passages from boiler to column and from column to glass are clear.

3. Close drain valve on glass and open drain valve on column. If steam blows freely from column drain, the column itself is clear.

4. Close column drain valve and open bottom valves to column and glass. Note whether water rises quickly to correct level. If action is

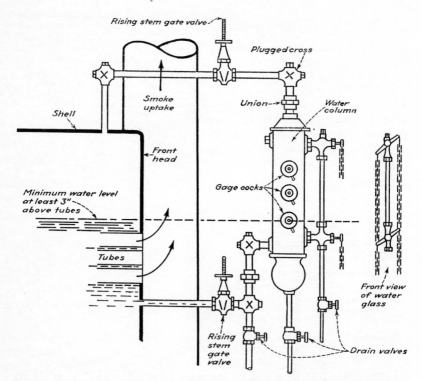

FIG. 8-5 Water column details shown for older hrt-boiler were vital for safe operation.

sluggish, some obstruction may still be in pipes or valves. Make sure all drain valves are tightly closed and all other valves wide open. Seal valves on water-column pipe connections in open position.

Q What points must be observed in fitting up a water gage?

A The gage must be well lighted and so placed that water level can be easily seen at all times. Do not use water glasses or water-glass guards that obscure water level. Automatic ball-check shutoff valves may be used if they comply with the ASME code. Quick-closing valves with lever handles and hanging chains shut off steam and water connections without danger of the operator's being scalded if a glass breaks. It is important to set visible bottom of gage glass at the correct distance above highest point of the boiler heating surface that might be damaged by low water.

On very high boilers the water-gage glass is sometimes set with its top tilted outward so that it may be more easily seen, and special forms of gages are sometimes used with lamps and mirrors to project gage readings to floor level. Flat glasses are used in gages for very high steam pressures and may be constructed to make water appear black and steam white.

In one make of high-pressure water gage, water space appears green in color and steam space red. These color changes make it possible to read the gage more easily.

Q What are gage cocks and where are they used?

A Gage cocks are small globe valves with side outlets and wheel or lever handles. They are a check on the water gage or a temporary means of finding water level when a gage glass breaks by showing whether water or steam blows out when a cock is opened. When the gage glass is placed directly on the boiler head or shell, gage cocks are also directly attached. When a water column is used, gage cocks are placed on the side of the column, as in Fig. 8-5.

According to the ASME code, each boiler must have three or more gage cocks located within the visible length of the water glass, except when a boiler has two water glasses independently connected to the boiler at least 2 ft apart. Firebox or water-leg boilers with not more than 50-sq-ft heating surface need have only two gage cocks. The bottom gage cock is placed level with the visible bottom of the water glass; others are spaced vertically at suitable distances.

Q Sketch two types of gage cock.

A Figure 8-6a is a simple gage cock with a wheel handle for opening and closing, while Fig. 8-5b is a spring-lever-handle type in which the valve is kept closed by a strong spring and opened by a chain-operated lever.

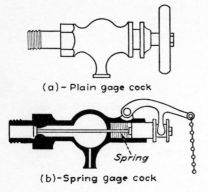

(a) - Plain gage cock

(b) - Spring gage cock

Spring

FIG. 8-6 Two types of gage cocks.

SAFETY VALVES

Q What is the function of the boiler safety valve?

A It prevents boiler pressure from rising above a certain predetermined pressure by opening to allow excess steam to escape into the atmosphere when that point is reached, thus guarding against a possible explosion through excessive pressure.

If the boiler has more than 500 sq ft of heating surface it should have two or more safety valves. In any case, safety-valve capacity must be such as to discharge all the steam the boiler can generate without allowing pressure to rise more than 6 percent above highest pressure at which any valve is set, and in no case more than 6 percent above the maximum allowable working pressure. All safety valves must be the direct spring-loaded pop type.

Q Explain the operation of the spring-loaded pop safety valve.

A It opens suddenly with a popping sound when the steam pressure reaches the point at which the valve is set to blow, hence the name "pop." This quick action is obtained in all pop valves by some method of suddenly increasing the upward force on the spring that holds the valve down on its seat. For example, on some types of safety valves, an extension or lip on the valve presents a greater area to the upward pressure of escaping steam as soon as the valve opens.

It is held open until the pressure drops a few pounds. Then it closes just as suddenly as it opened. This quick opening-and-closing action prevents "wiredrawing" of the steam and probable grooving of valve and seat.

Q What would cause a safety valve to stick to its seat?

A Most likely corrosion of valve and valve seat as a result of the valve's not being opened for a long period. To avoid this most dangerous condition, allow the valve to blow off periodically by using the hand lever to raise it from its seat or, preferably, by raising the steam pressure to the popping point.

Q Describe and sketch a spring-loaded pop safety valve.

A A common cast-iron-body type is shown in Fig. 8-7. The valve disk is held firmly on its seat by a heavy coil spring. The point at which the valve lifts and relieves the pressure is set by screwing the adjusting nut up or down and so decreasing or increasing compression of the spring. A lock nut keeps the adjusting nut from moving once it is set. The cap on top of valve may be sealed, if desired, to prevent access to the adjusting nuts. A hand lever lifts the valve from its seat if it is thought to be sticking, and a blowdown adjusting arrangement regulates number of pounds that the valve blows down before it closes. Figure 8-7 shows an enclosed spring.

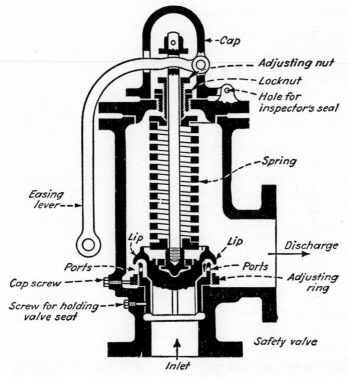

FIG. 8-7 Pop type safety valve for steam boiler must be thoroughly familiar to all operators.

An exposed spring is used when high steam temperatures might affect proper operation of the spring.

Q What is meant by *blowdown* of a safety valve, and how is it regulated?
A Difference between the pressure at which the valve pops and that at which it closes is the *blowdown*. The valve remains open until the pressure falls a few pounds below the original popping point because when it does open, a greater area of disk surface is presented to the upward rush of escaping steam and further upward pressure is exerted on the valve lip. In Fig. 8-7 the amount of blowdown may be varied by raising or lowering the adjusting ring on the outside of the valve seat, thus decreasing or increasing the area of the small ports through which the steam escapes from the "huddling" chamber.

The ASME code requires blowdown adjustment to be made in the factory and sealed. Changes which are required in the pressure at which the valve pops or in the amount of blowdown should be made by the boiler inspector or another competent person.

Q What is the reason for allowing blowdown on a safety valve?
A It prevents "chattering," that is, the rapid opening and closing of the valve that would take place if the steam were allowed to escape freely and there were no upward pressure on the spring other than that exerted on the bottom of the valve disk.

Q What are some points to consider in attaching a safety valve to a steam boiler?
A Connect the valve as close to the boiler as possible with no unnecessary pipe or fitting between. No other valve of any description may be placed between boiler and safety valve, nor on any escape pipe from it to the atmosphere. If a discharge pipe is used, it must have an open-ended drain to prevent water from lodging above the valve. Locate all safety valves so that the discharge is clear of running boards, working platforms, or any other place where escaping steam is likely to scald anyone. Connect every safety valve in an upright position with the spindle vertical. The valve must not be connected to an internal pipe and the hole in the boiler shell must be at least equal in area to the combined areas of all safety valves connected to the boiler. A muffler's construction must be such that it puts no back pressure on the valve.

Q How is the capacity of a safety valve tested?
A Tests are made under steam in the factory to determine the capacity, and the procedure to be followed in these tests is laid down in the ASME Code for Power Boilers.

If a test is required to be made on a boiler in operation, there are several ways in which this may be carried out. Probably the best check on safety-valve capacity is by an accumulation test. This is made by shutting off all steam outlets from the boiler and forcing the fires to the maximum. Under these conditions, the safety valve or valves should prevent the pressure from rising more than 6 percent above the highest pressure at which the valve or valves are set to pop and in no case more than 6 percent above the maximum allowable working pressure.

Q What is total upward force on a valve disk, 3 in. in diameter, when boiler steam pressure is 150 psi?
A Area of valve = diameter squared × 0.785
$$= 3 \times 3 \times 0.785 = 7.07 \text{ sq in.}$$
Total force = area of valve × pressure, psi
$$= 7.07 \times 150 = 1,060 \text{ lb} \quad \textit{Ans.}$$

Q If downward force exerted by the spring of a safety valve is 850 lb and diameter of the valve disk is 2½ in., at what pressure will the valve open?
A Area of valve disk = 2.5 × 2.5 × 0.785 = 4.91 sq in.
Opening steam pressure = spring pressure ÷ valve area
$$= 850 \div 4.91 = 173 \text{ psi} \quad \textit{Ans.}$$

Q Describe the high- and low-water alarm shown in Fig. 8-8.
A This form of high- and low-water alarm has a whistle placed on the water column and operated by rods and levers attached to floats or weights within the column. Movement of these weights or floats as the water level falls below or rises above the danger points causes the warning whistle to blow.

Figure 8-8 shows a high- and low-water alarm of the solid-weight type. It depends for its action upon the well-known fact that a body weighs less in water than in air. Upper weight is in the steam space, and lower weight is wholly submerged in water. Water in this design is about at normal operating level. In this position the weights are balanced and the whistle valve is closed.

If the level falls to a point where the lower weight is no longer submerged, the balance is disturbed. Lower weight, now unsupported, moves downward. This causes a corresponding upward movement of the upper weight, which blows the warning whistle.

If the water rises so high that the upper weight is submerged, its downward pull is lessened and the balance disturbed, so that the lower weight again moves downward and the upper weight rises, blowing the warning whistle as before.

The rods by which the weights are suspended are so attached to the short cross levers at the top that upper and lower weights must always move in opposite directions. Either low or high water moves the weights farther apart, and the resulting turn of the levers opens the steam valve to the whistle.

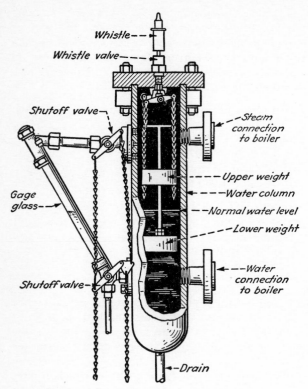

FIG. 8-8 High- and low-water alarm with gage glass at angle.

Q What is a fusible plug?

A This device protects the boiler from damage through low water. It is a bronze plug having a fine thread on the outside and a tapered hole in the center filled with a soft metal alloy, which is almost pure tin with a melting point of around 450°F.

Plug is screwed into the boiler-plate at some specified point above the highest point of the boiler heating surface that might be damaged by low water. In normal operation, one end of the plug is always covered by water while the other end is exposed to fire or hot gas. The tin filling does not melt as long as the plug is submerged, but if falling water level uncovers the plug, the tin melts and steam blows out of the central hole, thus warning of low water.

A fusible plug is not practicable in many modern boilers where pressures and temperatures are high and shells of drums are often thick, but it is still a valuable accessory in low- and medium-pressure boilers, especially in fire-tube ones. The ASME code gives correct location of the fusible plug in most well-known types of boilers.

Q Sketch and describe some approved forms of fusible plugs.
A Those recommended in the ASME code are shown in Fig. 8-9. The water-side plug (*a*) is screwed into the plate from the water side of the boiler; (*b*) shows a fire-side plug, so called because it screws into the plate from the fire side; (*c*) is a fire-side plug with the end recessed for a plug wrench to reduce the amount projecting beyond the plate surface. Plugs (*a*) and (*b*) have hexagonal ends and are screwed in with an ordinary wrench. The plug must project at least ³⁄₄ in. beyond the plate on the water side and should not project more than 1 in. on the fire side. The boiler pressure is always on the large end of the tapered soft-metal core.

Q What care must a fusible plug receive?
A Examine it every time the boiler is shut down for washing or inspection and clean all soot and scale off the ends. It must be renewed at least once a year.

Q How would you refill a fusible plug?
A The ASME code recommends that used casings not be refilled. But if it is necessary to refill a plug, heat it to melt out the old filling, clean the hole well, and tin it. Next, place a small plug of putty in the small end of the hole, set the plug on a smooth, level surface with the small end of the tapered hole down, and pour in some molten tin. When the plug cools, smooth off the end with a file, and it is again ready for use.

When melting the tin, take care not to overheat it.

Q Where should the fusible plug be placed in the following boilers: (1) hrt, (2) water-leg, (3) horizontal-drum water-tube?
A (1) Hrt boiler: in the rear head not less than 1 in. above the upper row of tubes. Measurement is taken from the upper surface of the tubes to center of plug; (2) water-leg boiler: in the highest part of the crown sheet; (3) horizontal-drum water-tube boiler: in the upper drum not less

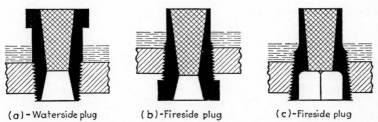

(a)-Waterside plug   (b)-Fireside plug   (c)-Fireside plug

FIG. 8-9 Three types of fusible plugs used for steam boilers.

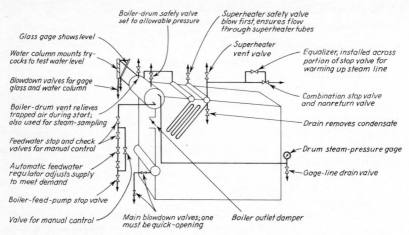

Glass gage shows level

Water column mounts try-cocks to test water level

Blowdown valves for gage glass and water column

Boiler-drum vent relieves trapped air during start; also used for steam-sampling

Feedwater stop and check valves for manual control

Automatic feedwater regulator adjusts supply to meet demand

Boiler-feed-pump stop valve

Valve for manual control

Boiler-drum safety valve set to allowable pressure

Superheater safety valve blow first, ensures flow through superheater tubes

Superheater vent valve

Equalizer, installed across portion of stop valve for warming up steam line

Combination stop valve and nonreturn valve

Drain removes condensate

Drum steam-pressure gage

Gage-line drain valve

Main blowdown valves; one must be quick-opening

Boiler outlet damper

FIG. 8-10   Trim serves to control flow of feedwater and steam, relieve excess pressure, and perform many other functions.

than 6 in. above the bottom of the drum over the first pass of the products of combustion.

Q What should be done if the feed-water supply to the boilers is suddenly shut off through injector, pump, or feed-line trouble?

A If the cause of the trouble is not immediately apparent, or is such that it will take some time to remedy it, firing must be stopped at once, before the water can reach a dangerously low level.

  If gas, oil, or pulverized fuel is being used, shut off the fuel supply and reduce the draft. If boilers are hand-fired with stationary grates, smother the fire with ashes, sand, earth, or fresh coal and reduce draft. Even with dumping grates it may be preferable to smother the fire rather than dump it in the ashpit and perhaps warp the grates. If boilers are stoker-fired, increase the stoker speed and feed to run out the live fire, cover the grates with green coal, and reduce the draft.

Q Is there other boiler trim in addition to that covered so far in this chapter?

A Figure 8-10 shows all the trim that serves to control flow of feedwater and steam, relieve excess pressure, remove unwanted condensation and solids, and indicate water level, temperature, and pressure in a boiler. For safe operation, firemen and stationary engineers must be familiar with each one and must know how to test them and maintain them in good working order.

COMBUSTION CONTROLS

Q Explain control systems for modern packaged boilers.

A Such systems generally include safety shutoff valves, limit switches, equipment for ignition proving, combustion control, and flame failure, plus start-up and shutdown programming. The control system senses and makes the boiler respond to changes in steam demand to the point of shutting the unit down when steam needs drop to zero. And if an unsafe condition arises, the boiler locks out.

Automatic regulation of fuel feed, in the proper ratio to the air supply, is the basic job for combustion control equipment. Regulation is initiated in response to a steam-pressure change.

Q What are the three basic combustion control systems?

A (1) On-off, (2) positioning, and (3) metering. An *on-off* system, Fig. 8-11, works between two steam pressure levels to start and stop both air supply and fuel feed. The simplest control, it has a minor drawback in that during the off period, natural draft through furnace carries away heat, lowering operation efficiency.

In a *positioning* system, Fig. 8-12, response to steam pressure is based on the assumption that, for example, a given damper position will provide sufficient air for a given fuel flow to hold a constant fuel-air ratio throughout the load range. Yet, at any position, air flow may be influenced by voltage variations at the blower motor and by draft loss through the boiler caused by soot or slag and other variables. Thus, although a positioning system may hold steam pressure constant, combustion efficiency may not stay at a high level.

A *metering-type* control, Fig. 8-13, goes a step beyond the positioning setup by actually measuring fuel and air flows and draft loss. These val-

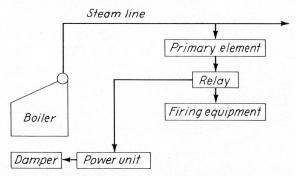

FIG. 8-11 On-off control system has low load range.

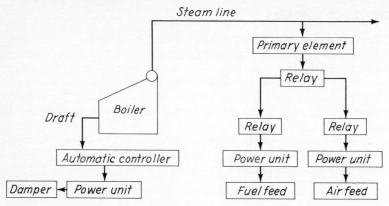

FIG. 8-12 Positioning type has fuel and air tied together in predetermined ratio.

ues are balanced against signals for more or less steam pressure. The result is compensation for variables which tend to influence the positioning system's operation.

Briefly, metering control yields top precision, and this higher precision is important when a boiler is operated under fluctuating loads for long periods. Generally, both positioning and metering systems include an on-off feature, extending the below-load range beyond that of modulating.

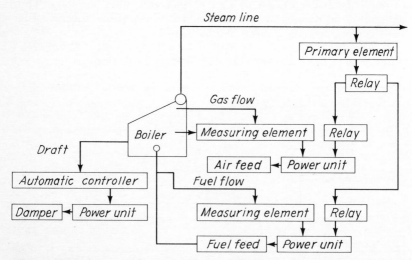

FIG. 8-13 Metering control measures fuel and air flow against signals that indicate more, or less, steam pressure.

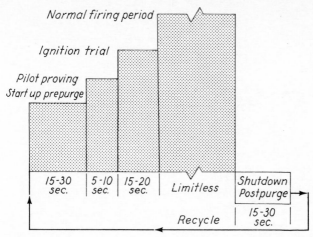

FIG. 8-14 Programming-sequence safeguards boiler throughout operating cycle.

Today's combustion controls are available with hydraulic, electric, or pneumatic relays to transmit impulses to control points.

Q Explain programming sequence.
A Programming-sequence, Fig. 8-14, safeguards the boiler throughout the operating cycle. On start-up, purge operation forces air through the furnace and gas passes to clear out any combustible gas pockets. Ignition is next. Depending on the fuel, the ignition system may be electric, spark, light-oil-electric (in which a spark ignites a thin oil which in turn touches off the main fuel), or gas-electric (spark ignites a gas pilot).

A flame scanner is on the job during ignition. The control system won't let the main gas or fuel-oil valve open unless the scanner indicates ignition is normal (ignition proving). Even after lightoff, the scanner continues to supervise the burner. If flame is lost, the scanner shuts the unit down. Following shutdown, the blower again purges the furnace and gas passes before the ignition cycle recurs.

LOW-WATER CUTOFFS

Q Explain the low-water cutoff.
A This device is separate from the programming sequence control, and is extremely important to general safety. Its job is to shut down the boiler immediately if water drops to a dangerously low level. There are three basic types, Fig. 8-15. In (a), the *float-magnet* design, a nonferrous sleeve encloses a ferrous plunger on a float rod which stays out of the perma-

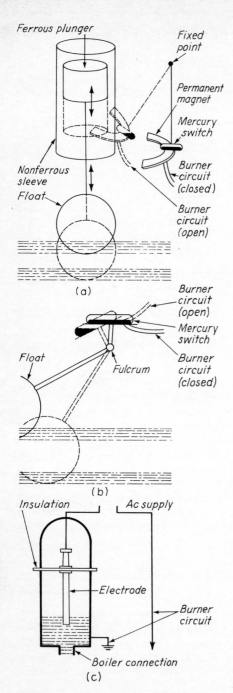

FIG. 8-15 Basic low-water cutoff types. (a) Float-magnet; (b) float-linkage; (c) electrode-probe.

nent magnet's field when the water level is normal. A switch on the magnet holds the burner circuit closed. When lowered water drops the float, the magnet swings in to the plunger, which tilts the mercury switch and opens the burner circuit.

In the *float-linkage* type (*b*), the float is connected through a linkage to a plate supporting a mercury switch. On a horizontal plane when the water level is normal, the switch holds the burner circuit closed. When lowered water drops the float, the linkage tilts the plate and attached mercury switch, opening the burner circuit.

In the *electrode-probe* type (*c*), an electrode, insulated from its grounded housing, is partially submerged when the water level is normal. With the burner circuit in series, current flows from the electrode through the water to the housing to hold the burner circuit closed. Lowered water bares the tip, breaking the circuit to the burner.

Q Explain the location of the low-water cutoff, Fig. 8-16, for heating boilers.

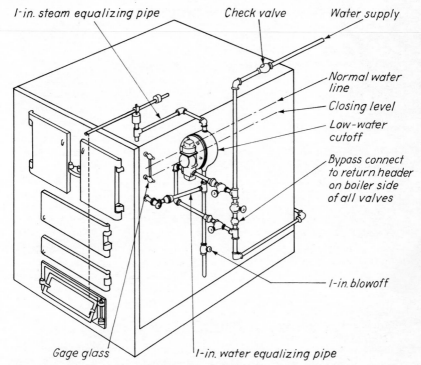

FIG. 8-16 Low-water cutoff correctly installed has water and steam space-equalizing lines connected properly to get true boiler water level in float chamber.

A Whether the unit is a steel boiler with independent water columns, or a cast-iron boiler with a water glass in the first section, it's important to connect the water space-equalizing line at a point about 6 in. below the water glass (or above the boiler's firebox) where there is a large volume of water and slow circulation.

Q Water column gage glasses are hard to read from a distance, especially if the glass tube is not clean and the water not clear. Explain one way to solve the problem.

A Figure 8-17 shows a gage glass with a strip in back, painted with black and white angles. The water level stands out because the stripes run in opposite directions starting at water level. The amount of strip distortion through the water-filled tube section depends on tube size and distance from tube to painted card. A little experimenting soon shows the best strip width and distance from glass.

FIG. 8-17 Sure-fire water-level finder.

FIRE SAFETY SWITCH

Q What is the purpose of the ignitable cord on the fire safety switch, found on many horizontal rotary atomizing oil burners?

A These burners have as one of the protection controls a fire safety switch which shuts down the entire system instantly in the event of a flareback. The switch is held closed by an ignitable cord which will burn quickly, thus releasing the switch, which snaps open and breaks the electrical circuit, thereby shutting down the burner.

CAUTION: Do *not* replace this cord with a wire, or any nonburnable cord, as is often done.

NOTE: We show in Table 8-1 a list of accidents in 1973 to power and heating boilers and to unfired pressure vessels. All persons working with boilers should study the types and incidence of accidents that occur most often, then study their equipment and set up a system of inspection and operation to prevent such accidents to units under their care.

WHY BOILERS EXPLODE

Q With the many safety devices and automatic controls on present-day boilers, why are there so many boiler failures?

TABLE 8-1 Accidents Reported by National Board Members and Other Authorized Inspection Agencies (For the period January 1, 1973–December 31, 1973)

Type	Accidents	Injuries	Deaths
Power Boilers:			
Rupture of drum during hydrostatic test	6	16	0
Shell rupture	11	4	0
Furnace explosions	57	11	2
Flarebacks	16	3	0
Low water failures	44	4	0
Miscellaneous overheating failures	183	1	0
Piping failures	73	6	0
Poor maintenance of controls	49	0	0
Unsafe practice	48	1	0
Construction Code violations (welds)	2	0	0
Dry fired	462	0	0
Total	951	46	2
Heating Boilers:			
Dry fired	47	0	0
Gas explosion	11	0	0
Shell rupture	6	0	0
Furnace explosions	47	43	0
Flarebacks	2	2	0
Low water failures	78	0	0
Runaway burners	6	0	0
Inadequate safety valves	8	2	0
Construction Code violations	2	0	0
Cast iron boilers	843*	0	0
Domestic hot-water heaters	7	4	0
Total	1,057	51	0
Unfired Pressure Vessels:			
Nitrogen tank	0	0	0
LPG vessels—in service	1	3	0
LPG vessels—filling and handling	3	106	12
Piping and hose failures	16	0	1
Air tank explosions	21	2	0
Air and water (or oil)	2	0	0
Jacketed kettles	8	0	1
Improperly vented tank or heat exchanger	1	0	0
Quick-opening door latch failures	5	0	1
Miscellaneous UPV failures	30	3	0
Brittle fracture during shop hydrostatic test	2	0	0
Unsafe practice	117	19	8
Construction Code violations	2	0	0
Safety valve failures	9	4	1
Total	217	137	24
GRAND TOTAL	2,225	234	26

* Caused by dry firing, lack of preventive maintenance, lack of expansion provisions, etc.

A The main reason is the lowering or complete absence of operating standards. Too many owners have only one thought in mind when buying boilers provided with automatic controls—they hope to operate without the benefit of qualified operators. Until such time as all interested parties (these include government and insurance company inspectors, boiler manufacturers, code-making organizations, and engineers) can convince boiler owners that they cannot shed all responsibilities toward safe operating practices by substituting automatic devices, boiler accidents will continue at the present alarming rate.

Q What is the most common type of boiler accident occurring today?
A The greatest accident producer today is low water, which usually results in overheating and loosening of tubes, collapse of furnaces and, in some cases, complete destruction of the boiler. In certain classes of boilers a low-water condition can set the stage for a disastrous explosion that can cause serious loss.

Q Why is low water responsible for most boiler accidents in the past decade?
A Primarily because of a lack of boiler-operating and maintenance standards. This is true despite the fact that most boilers, especially heating ones, are provided with automatic feed devices as one of the many automatic controls. But it is these devices that lull the owners into a false sense of security. Many owners feel that the boiler is fully automatic and completely protected from accidents. Not understanding fully how potentially dangerous a fired pressure vessel can be, the owner (or anyone else) does not seem to take the slightest interest after installing one of these so-called automatic boilers.

The fact is that automatic feed devices, like all other automatic devices, will work perhaps a thousand times, perhaps many thousand times. But at some time they will probably fail, usually with disastrous results. That is why it's the duty of everyone concerned with boilers to realize that unless proper operating and maintenance standards are instituted, accidents are certain to occur.

Q Isn't it true that, in practically all cases, low-water fuel cutouts are installed on all automatically fired boilers to protect the boiler from overheating, in the event of failure of the feeding device? If such is the case, what causes the failure?
A Practically all boilers today are provided with low-water fuel cutouts. What most people don't realize is that in most accidents, regardless of the type of cutouts, a *series* of failures occurs. Thus, the basic fault may be failure of the automatic feed device to operate. Then we experience failure of the low-water fuel cutout. The net result is overheating and

burning of the boiler metal. The failure of the automatic feed device and the subsequent failure of the low-water fuel cutout to operate stem from the same basic cause—lack of operating and maintenance standards on the part of the owner.

Q How should a low-water cutout be tested?
A The only positive method of testing a LWCO is by duplicating an actual low-water condition. This is done by draining the boiler slowly while under pressure. Draining is done through the boiler blowdown line. We find that many heating boilers are not provided with facilities for proper draining—an important consideration.

Many operators mistakenly feel that draining the float chamber of the cutout is the proper test. But this particular drain line is only provided for blowing out sediment that may collect in the float chamber. In most cases the float will drop when this drain is opened because of the sudden rush of water from the float chamber. Every boiler inspector can tell you of numerous experiences of draining the float chamber and having the cutout perform satisfactorily. But when proper testing was done by draining the boiler, the cutout failed to function.

Q What percentage of boiler losses are caused by low water?
A Approximately 75 percent.

Q Various articles have been published on the subject of safety valves, particularly the low-pressure type with a setting of only 15 psi. If it is true that all boilers furnished today are provided with ASME-approved valves, why should anyone take exception to them and why shouldn't they work?
A In the first place, most people have a misconception of the term *ASME-approved*. To set the record straight, the ASME itself does not approve a type of safety valve. By referring to the ASME code, Section I–Power Boilers and Section IV–Low Pressure Heating Boilers, you will see that both codes contain limited design criteria. The codes also require a manufacturer to submit valves for testing. Such tests are solely for pressure-setting and relieving capacity.

In brief, the ASME symbol on a safety valve attests to the fact that the limited design criteria and the materials outlined in the code have been supplied by the manufacturer and that the relieving capacity and set pressure stamped on the valve have been proved.

Getting back to the first part of the question, in some cases experience indicates that a particular type of safety valve has an inherent design weakness. After a short period of operation, the disk may be subject to sticking due to close clearances. This condition will render the valve useless, and the boiler will be without the benefit of overpressure protection.

In regard to the second part of the question, the failure of safety valves to work is usually caused by build-up of foreign deposits that result in "freezing." This is an indication that the valve has not been regularly tested or examined. One of the greatest causes of foreign deposit buildup is a "weeping" or leaking condition. The only way to be sure that a valve is in proper operating condition is to set up and adhere to a regular program of testing the valves by hand while the boiler is under pressure. Also, any weeping or leaking valve should be immediately replaced or repaired. This is important.

Q How often and in what manner should boiler safety valves be tested to make certain they are in proper working order?
A Low-pressure (15 psi) safety valves should be lifted at least once a month while the boiler is under steam pressure. The valve should be opened fully and the try lever released, so the valve will snap closed. For boilers operating between 16 and 225 psi, the safety valves should be tested weekly by lifting the valves by hand. On these higher-pressure boilers it is good practice to test the safety valves by raising the pressure on the boiler. This can usually be done when the boiler is being taken off the line. Then, if the valve "feathers" from improper seating, it can be corrected when the boiler is cold.

Q What can be done to prevent boiler failures?
A Boiler failures can at least be greatly reduced if boilers are placed under the custody of properly trained operators. This means that boiler owners must use sound judgment when employing boiler operators. Everyone with an interest in boilers must be encouraged in educating boiler owners and operators in proper operating procedures.

A very important step is establishing a regular program for testing of controls and safety devices, then faithfully following through.

Further, a program must be established for periodic maintenance of controls and safety devices. First thing to realize is that providing boilers with the most modern proved controls and safety devices is no guarantee that you will not have boiler failures. Any control or safety device is only as good as the testing and maintenance it receives. You cannot ever relax on these two.

Q What can a boiler owner do to assure that he has taken all possible steps to prevent boiler failure?
A First, purchase the best equipment available for a given service. Second, make certain that the boiler is properly installed and equipped with all the necessary appurtenances and safety devices. Third, before taking final acceptance, specify that the installation be inspected by a commissioned inspector in the employ of an insurance company or the state or

municipality. By doing so, the owner will be assured that the equipment and the installation meet the legal requirements of the particular state or municipality. Fourth, the owner should provide the operator with a log book and a set of preventive maintenance and testing procedures. He should insist that such procedures be followed religiously and that the results of the test and maintenance be recorded and be made a permanent part of the boiler room log.

SUGGESTED READING

REPRINT SOLD BY *Power* MAGAZINE,
  Power Handbook, 64 pp.

BOOKS
  Elonka, Stephen M., and Anthony L. Kohan: "Standard Boiler Operators' Questions & Answers," McGraw-Hill Book Company, New York, 1969.
  Elonka, Stephen M., and Julian L. Bernstein: "Standard Electronics Questions & Answers," vols. I and II, McGraw-Hill Book Company, New York, 1964.
  Elonka, Stephen M., and Alonzo R. Parsons: "Standard Instrumentation Questions and Answers," McGraw-Hill Book Company, New York, 1962.
  Elonka, Stephen M., and Joseph F. Robinson: "Standard Plant Operators' Questions and Answers," vols. I and II, McGraw-Hill Book Company, New York, 1959.
  Elonka, Stephen M.: "Standard Plant Operators' Manual," McGraw-Hill Book Company, New York, 1975.

9

FEED AND BLOWOFF ACCESSORIES

Automatic feed-water regulators, blowoff valves, and related fittings do their job unfailingly *only* if the operator is familiar with them, uses them as intended, and keeps them in perfect working condition. Even simple valves used in boiler room piping systems cannot be used for just any service, as their marking indicates. Here we explain the proper uses of these fittings.

It is important to remember, incidentally, that boilers may not always be blown down and discharged into the sewer, since some local ordinances forbid the practice. That is why blowoff tanks are used.

Operators who are proud of their licenses and boiler rooms usually run a "tight ship" and keep their equipment "shipshape"—provided, of course, that management cooperates.

FEED LINE

Q Where should feedwater be introduced into the boiler?

A It should be introduced at a point where it will not discharge directly against any riveted joint or surface exposed to high-temperature gas or direct radiation from the fire. When feedwater must be introduced close to riveted joints or heating surface, use an internal baffle plate to deflect the entering water from the joint or heating surface. In some boilers internal pipes lead the water to the most suitable point of discharge inside the boiler. Some fire-tube and water-tube boilers have internal feed pipes for this purpose. Actual discharge of feedwater should always be below the minimum water level.

Q What does the ASME Code for Power Boilers say regarding means of supplying feedwater?
A Two means of supplying feedwater to the boiler should be provided for all boilers having more than 500 sq ft of water-heating surface. These may be two pumps, two inspirators, two injectors, or any suitable combination. If another source of supply is available at a pressure at least 6 percent higher than that at which the safety valve is set, it may be considered as one means of supply.

Weld

FIG. 9-1   Pad for feed-pipe connection to boiler shell.

Q How should the feed pipe be attached to the boiler shell?
A Figure 9-1 shows an approved form of threaded flange connection. This provides for both internal and external feed pipes. Both pipes should be so threaded that they will make tight joints without actually butting the ends together.

Q What valves are necessary on the boiler feed-water line at the boiler shell, and what are their functions?
A A stop valve must be placed next to the boiler shell so that the entire line can be shut off from the boiler during any emergency work on the check valve without shutting the boiler down. At all other times the stop valve is left wide open. The check valve is placed next to the stop valve. It is so constructed that the feedwater passes through it freely into the boiler, but, if feed-water pressure falls below boiler pressure, the check valve closes and prevents steam or water from the boiler backing up into the feed line.

Q Describe an arrangement of feed-water connections to a fire-tube boiler.
A The connection in Fig. 9-2 consists of a swing-check valve, gate-stop valve, cross with brass plugs, flanged and threaded pad, and internal feed pipe. A *gate* valve is superior to a *globe* valve in that it provides a *straightway* opening, but the seats and disks are more easily renewed in the globe valve. If a globe valve is used here, connect it so that the inlet is

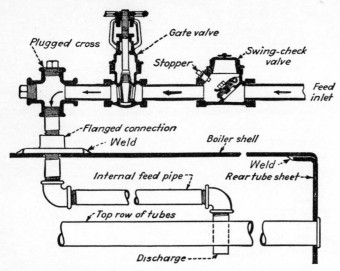

FIG. 9-2   Feed-water connections for fire-tube boiler.

under the valve disk. Then, if the valve stem breaks, the feed-water pressure will hold the valve open and allow water to enter the boiler.

The swing-check valve is simply a hinged disk that swings open as the feedwater forces its way through and drops back on its seat when feedwater pressure falls below boiler pressure. The swing-check valve also gives a straightway passage.

AUTOMATIC FEED-WATER REGULATORS _____

Q What is an automatic feed-water regulator?
A This device automatically regulates feed-water supply according to load, and so does away with hand operation of valves on feed lines. Most feed-water regulators are controlled by temperature, their action depending upon expansion and contraction of some metal part of variation in vapor pressure.

Q Describe some type of automatic feed-water regulator operated by expansion and contraction of metal.
A Figure 9-3 diagrams a feed-water regulator operated by expansion and contraction of an inclined metal tube, or thermostat in line with the water gage, and connected to the steam and water spaces, as shown. One end of tube is fixed; the other end is free to move. The free end operates

the control valve in the feed-water line through a system of levers. The water level rises and falls in the expansion tube just as it does in the boiler.

The upper end of the tube, being filled with steam, is at approximately steam temperature, but the lower end, which is full of water, is but little warmer than the boiler room. With boiler water at half glass, the tube will also be half full of water, as shown at 2, the control valve in the feed line being half open, as at 2'.

When the load on the boiler increases, the water level falls, exposing more of the tube to steam. This expands the tube, opening the control valve to position 1'. When the load decreases and the water level rises, the tube contracts, moving the control lever to 3', thus closing the control valve, which is balanced to operate with minimum effort.

Q Describe a type of automatic feed-water regulator that is operated by changes in vapor pressure.

A The regulator shown diagrammatically in Fig. 9-4 has an inclined tube set in line with the water gage and connected top and bottom to the steam and water spaces of the boiler. This tube is enclosed in an outer casing, which is in turn connected by piping to a metal bellows attached to the feed-water control valve. The outer casing has a great number of fins for the purpose of dissipating heat rapidly by radiation and it, together with the connecting pipe and bellows, forms a closed low-pressure system with steam in the upper part of the casing and water in the connecting pipe and bellows.

In the position shown, if load is constant, the feed-water control valve is open just enough to keep a steady water level in the boiler. If the steam demand increases, the water level will fall in the boiler and in the upper inclined tube, exposing a greater length of the tube to steam. The tem-

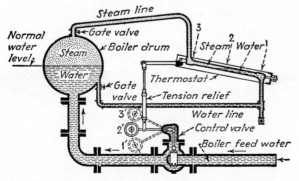

FIG. 9-3 Thermostatic automatic feed-water regulator.

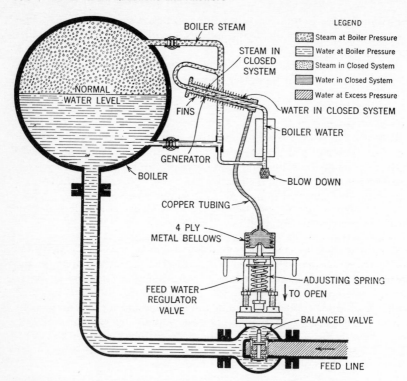

FIG. 9-4 Vapor-pressure-operated automatic feed-water regulator.

perature of the tube will then rise and this will generate more steam in the outer closed circuit, with a consequent rise in pressure. This increased pressure acting upon the water will cause the bellows to expand against the upward pressure of the adjusting spring, opening the feed-water control valve, and allowing more water to enter the boiler.

If the boiler load decreases, the water level will rise in the boiler and in the inner inclined tube. The inner tube will cool down to some extent, and some of the steam in the outer casing will condense, causing a drop of pressure in the closed system. The upward pressure of the adjusting spring will now cause the bellows to contract and the valve stem will rise, partially closing the valve, and allowing less water to enter the boiler.

BOILER BLOWDOWN

Q What is the purpose of the blowoff valve?

A It has several functions. When necessary it empties the boiler for cleaning, inspection, or repair. It blows out mud, scale, or sediment when the boiler is in operation and prevents excessive concentration of soluble impurities in the boiler.

Q What grade of pipe and types of valves should be used on boiler blowoff lines?

A Use extra-heavy pipe and only valves that offer an unrestricted passage. Plug cocks, especially those that can be lubricated, and quick-opening valves are suitable, as well as certain other Y and angle slow-opening valves that are made for this purpose. Do not use ordinary globe valves on blowoff lines as they tend to prevent passage of mud or scale and cannot be cleared, when plugged, by passing a rod or heavy wire through the blowoff pipe and valve. Gate valves, although they offer a straightway passage, are unsuitable for blowoff because scale can lodge beneath the gate and prevent its closing.

The ASME code specifies that all boilers carrying over 100 psi working pressure, except those used for traction or portable purposes, shall have two blowoff valves on each blowoff pipe. These may be two slow-opening valves, or one slow-opening valve and one quick-opening valve or plug cock. Traction and portable boilers must have one slow- or quick-opening blowoff valve.

Q What is the proper sequence in opening and closing blowoff valves when blowing down a boiler?

A When a boiler is equipped with both a blowoff valve and cock or quick-opening valve, Fig. 9-5, in the same blowoff connection, always open the cock or quick-opening valve first and the blowoff valve second. To close, always close the blowoff valve first and the cock or quick-opening valve second. Caution the boiler operator to open and close

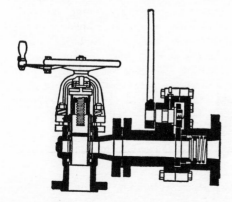

FIG. 9-5 Angle and quick-opening straightway blowoff valves.

blowoff valves and cocks slowly, to reduce shock as much as possible. And *never* take your hands off the blowoff valve while it is open, nor your eyes off the gage glass.

Q When and how often should boilers be blown down?

A Boilers should be blown down only during periods when steam production is at a minimum. The reason is that circulation in some boilers is very sensitive. Blowing down when steam production is at a maximum could upset the circulation so badly that serious damage could be done to some parts of the boiler, especially the tubes.

Blowdown valves on waterwalls serve primarily as drain valves. Never blow down waterwalls when the boiler is in operation. If difficulties arise that require blowing down waterwalls, do this only under banked conditions or only in accordance with the boiler manufacturer's instructions.

Q What are some typical blowoff valves?

A Figure 9-5 shows a typical blowoff arrangement: an angle valve in series with a quick-opening valve.

Figure 9-6 shows details of a plug cock. Any plug cock used for blowoff must have a guard or gland to hold the plug in place, and the end of the plug must be marked in line with the passage through the plug so that anyone can see whether the valve is open or closed.

Q When is a blowoff tank necessary, and how is it constructed?

A Necessary when there is no open space available into which blowoff from the boilers can discharge without danger of accident or damage to

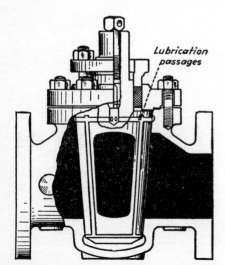

Lubrication passages

FIG. 9-6 Plug-cock blowoff.

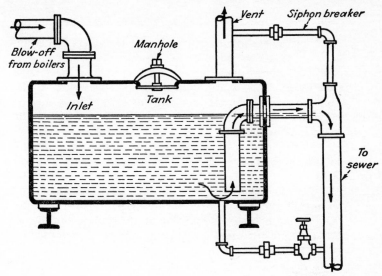

FIG. 9-7 Blowoff tank piping arrangement for reliable operation.

property. For example, where the blowoff must be discharged to a sewer, the sewer would probably be damaged by blowing hot water under high pressure directly into it, and water and steam might possibly back up into other sewer connections.

The blowoff tank always stands nearly full of water, as in Fig. 9-7. When the boilers are blown down, the cooler water in the bottom of the tank overflows to the sewer as it is displaced by the entering hot water. The large open vent prevents pressure from building up in the tank, and a small pipe runs from the top of the overflow into the vent pipe to prevent any siphoning action that might empty the tank.

Q List the precautions to take with blowoff valves.

A 1. Avoid accidental opening of a blowoff valve, especially of the quick-opening type. Guard against this by removing the handle or locking it in a closed position when the valve is not in use.

2. Open and close the valve slowly to prevent water hammer and possible rupture of pipes, valves, or fittings. Double blowoff valves protect against this trouble and avoid the need to shut down immediately if one of the valves fails to close properly.

3. If a blowoff valve appears to leak when closed, don't try to force it onto its seat, but open it again so that boiler pressure will wash the valve and remove whatever may be lodging on the seat. Forcing only damages the valve, whereas the extra blow-through will probably remove leakage cause.

VALVES FOR BOILERS

Q What types of stop valves should be used on steam boilers and steam mains?

A Fit each main or auxiliary discharge steam outlet, except safety valve and superheater connections, with a stop valve placed as close to the boiler as possible. When outlet size is over 2 in. (pipe size), valves must be the outside-screw-and-yoke type to indicate by the position of the spindle whether the valve is open or closed.

When two or more boilers are connected to a common steam main, the steam connection from each boiler having a manhole must have two stop valves in series, with an ample free-blowing drain between them, the discharge of the drain to be in full view of the operator when opening or closing the valves. Both valves may be of the outside-screw-and-yoke type, but it is preferable that one be an automatic nonreturn valve. This should be placed next to the boiler so that it can be examined and adjusted or repaired when the boiler is off the line.

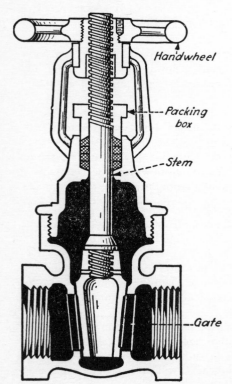

Handwheel

Packing box

Stem

Gate

FIG. 9-8 OS&Y gate valve.

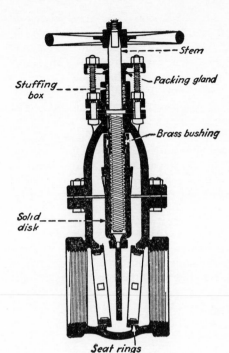

Stem

Stuffing
box

Packing gland

Brass bushing

Solid
disk

FIG. 9-9   Inside-screw gate valve.

Seat rings

Q Describe an *outside-screw-and-yoke* valve.

A Commonly called an OS&Y valve, Fig. 9-8, its screwed part of the valve stem works in a bushing *outside* the main valve body. This bushing is fastened solidly to the handwheel and turns in a socket at top of the outside yoke. The threaded valve stem can rise and fall but cannot turn. When the handwheel is turned, the threaded bushing also turns, thus raising or lowering the valve stem and opening or closing the valve. Very little of the threaded stem can be seen above the handwheel when the valve is closed, but a considerable length shows when the valve is open. Thus the amount of valve stem in sight above the handwheel is a positive indication of whether the valve is open or closed.

Q Describe an *inside-screw gate* valve.

A The stem of the valve shown in Fig. 9-9 rotates with the handwheel but does not rise or fall. Thus rotation screws the wedge up or down without any outside indication of the valve being open or closed.

Q Explain the construction and operation of an *automatic nonreturn* valve.

A This valve is usually placed on the main steam outlets of boilers in battery with others, in addition to the OS&Y stop valve. The automatic nonreturn valve shown in Fig. 9-10 can be closed by screwing down the outside stem but can only be opened by boiler-steam pressure, as the outside stem is not attached to the valve. The dashpot on top of the valve spindle cushions the valve movement and prevents chattering.

When a boiler is about ready to cut in, the OS&Y stop valve is opened. As soon as boiler pressure rises a little above the pressure in the steam main, it raises the nonreturn valve and automatically puts the boiler on the line. During operation, if for any reason the boiler pressure falls below the main header pressure, the nonreturn valve closes and cuts the boiler out. It can also be used in this way to cut out the boiler when it is being taken off the line for cleaning or repair. It really acts as a check valve, allowing steam flow from the boiler to the main but preventing steam flow from the main to the boiler.

Q What is the function of a *double-acting automatic nonreturn valve?*
A Like the single-acting nonreturn valve shown in Fig. 9-10, it allows steam to flow from the boiler to the main as long as boiler pressure is higher than pressure in the main. It closes when the boiler pressure falls

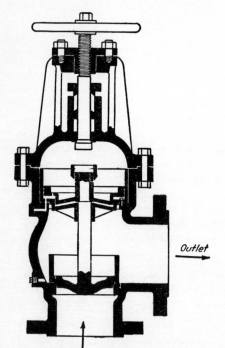

Outlet

FIG. 9-10 Automatic nonreturn valve.

From boiler

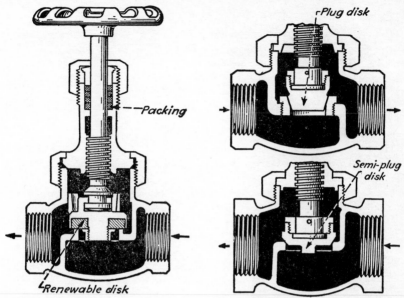

FIG. 9-11 Globe valve with renewable disk and flat seat.

FIG. 9-12 Globe valves with tapered disks and seats.

below the steam main pressure, but, in addition, it closes against the boiler pressure if the steam main pressure falls below a certain point. It thus protects against heavy steam flow from the main into the boiler in the event of some accident, such as the bursting of a tube. It also prevents a heavy flow of steam from the boiler into the main if some accident to the main or its connections suddenly lowers the steam pressure.

The double-acting automatic nonreturn valve can also cut the boiler on or off the line in the same way as the single-acting nonreturn valve.

Q What are the commonest types of valves for steam, water, air, or gas service?

A The two commonest are *gate* and *globe* valves. Gate valves give a straightway opening and are therefore suitable for pipe lines carrying heavy fluids, as they offer little resistance to the flow. Globe valves, so called because of their globular shape, do not give a straightway passage and are therefore not so suitable as gate valves where free and unrestricted passage is particularly desirable. However, globe are preferred to gate valves on most small pipe lines because disks and seats can easily be renewed. Sectional views of small globe valves are shown in Figs. 9-11 and 9-12. Two types of gate valve are shown in Fig. 9-13.

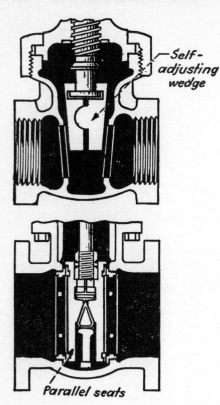

FIG. 9-13   Types of gate valves.

Q Sketch some form of *reducing* valve and describe its construction and operation.

A A *reducing* valve is simply an automatic throttle valve for use where low-pressure steam for heating or process is taken from high-pressure mains. Figure 9-14 shows a weighted-lever reducing valve for medium pressures. Referring to the figure, when there is no pressure on the low-pressure side, the weights are so adjusted as to keep the balanced valve open. As pressure builds up on the low-pressure side, it acts upward on the thin metal diaphragm attached to the valve stem, to lift the weight and close the valve. Any desired reduction in pressure within the limits of the valve is accomplished by adjusting the position of the weight on the lever.

Stop valves should be placed on both sides of the reducing valve to cut it out if it should get out of order. Run a bypass around the valve to provide a means of continuing the supply of steam temporarily. A pressure gage and a low-pressure safety valve should be placed on the low-pressure side.

Springs are used instead of levers on some makes of reducing valves. Plungers, operated by steam, air, oil, or water pressure, are also used to control the amount of valve opening in various forms of reducing and regulating valves.

Q What are automatically controlled valves?

A They are valves fitted with some means of opening and closing other than, or in addition to, the handwheel or lever. These may be a solenoid or electric motor, or compressed-air or hydraulically operated motors, or plungers. Very large valves that cannot be opened or closed easily by hand are generally equipped with power-operated mechanisms. Sometimes small valves also are fitted with an automatic device for opening and closing by remote control.

Q What does the marking *200 WOG* on a valve body mean?

A A body marked *200 WOG* (water, oil, gas) means the valve is safe for no more than 200 psi, whether for *cold* water, oil, or gas. A valve marked *125 S*, for example, means that the valve is for 125 psi saturated *steam,* at its equivalent temperature of 344°F (steam tables give the temperature). For cold service, this 125 S valve may be used up to 300 psi. This shows how dangerous it is to use a *cold* valve for *hot* service. Always check the valve body before installing valves in a system.

Q Explain metals used for boiler room valves.

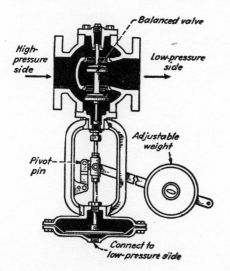

FIG. 9-14 Weighted-lever reducing valve.

A Metals for valves installed in boiler piping systems are usually brass, bronze, iron, or steel. Almost any of these metals is good for steam, cold water, oil, or gas. But each has certain limits. For example, never use brass above 500°F. Bronze can stand 350 psi pressure and temperature up to 700°F. Bronze and brass valves usually come in smaller sizes, up to and including 2 in. Above this size, use iron up to 250 psi and 500°F. Steel is for temperatures above 550°F, and is also used when the valve must withstand high internal pressure. Of course, metals for valves and piping in supercritical pressure systems (3,206.2 psi, 705.4°F) and for nuclear power plants are more specialized, as covered in the boiler code.

SUGGESTED READING

Reprints sold by *Power* magazine
  Power Handbook, 64 pp.
  Steam Generation, 48 pp.

Book
Elonka, Stephen M.: "Standard Plant Operators' Manual," McGraw-Hill Book Company, New York, 1975.

10

CONSTRUCTION, CLEANING, AND INSPECTION

Keeping a boiler clean and knowing *where* to look for defects are vital to safe, efficient operation. Because a knowledge of boiler construction is needed to evaluate the seriousness of a defect, here we cover some of the more critical components that make up a steam boiler. And to make sure the reader knows *what* to do when a serious defect is found, some details are spelled out in this chapter.

And how is a heating boiler protected when it is laid up for the summer season, or for any long period? Is it enough just to shut off the fuel, close all the valves, and let it wait for the next start-up? Here we cover lay-up of boilers so they are "raring to go" when needed at the next start-up — not badly corroded, leaking, and needing major repairs.

SHELL OPENINGS

Q What is the common form of manhole opening in flat surfaces?

A It is usually elliptical and flanged *inward* from solid plate, Fig. 10-1. When the plate edge is less than $^{11}/_{16}$ in. thick, the gasket-bearing surface is increased by shrinking a narrow ring around the flange and facing off both ring and plate edge to form a smooth joint.

Elliptical manholes must not be less than 10 by 16 in. or 11 by 15 in. inside. Circular manholes must not be less than 15 in. inside diameter.

Q What is the common form of manhole in curved surfaces?

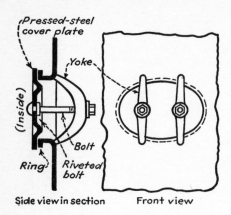

FIG. 10-1 Manhole in flat surface.

A The manhole is usually elliptical, with long axis at right angles to the axis of the shell, and is reinforced by a flanged ring welded to the shell, Fig. 10-2. A ring is also shrunk around the flange of the reinforcing ring if more gasket bearing surface is required.

Q How are manhole cover plates and yokes made, and what materials are used in their construction?
A Manhole cover plates and yokes are made from rolled, forged, or cast steel for pressures over 250 psi and temperatures above 450°F. Below this temperature and pressure they may be made from cast iron.

The cover plate in Fig. 10-1 is made from steel plate. The bolts are riveted to the plate, and the yokes or dogs are cast steel or pressed steel.

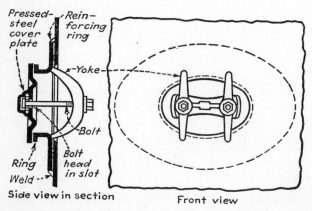

FIG. 10-2 Manhole and cover details in curved surface.

The cover plate, Fig. 10-2, is also pressed from steel plate, but the bolt heads fit loosely in slots instead of being riveted solidly to the plate. There is no possibility of leakage past the bolt heads with this construction, and probably less liability of bolt breakage, but the riveted-bolt manhole cover is more easily handled and does not often give any trouble.

Q Why are manholes and handholes usually elliptical in shape?

A The elliptical cover plate can easily pass through the manhole when it is necessary to take it out or insert it in place. This shape of manhole also affords easier access to the boiler with the minimum area of opening, and so does not weaken the plate to the same extent as openings of larger area. Figure 10-3 shows a handhole opening and handhole cover plate.

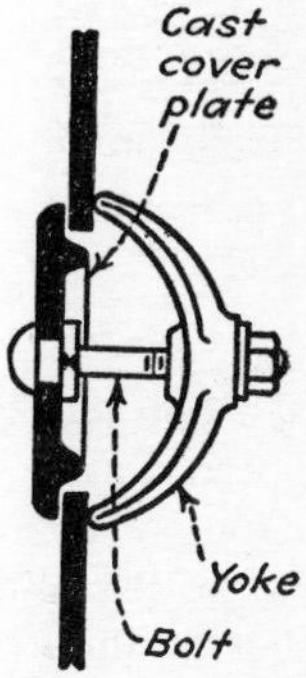

FIG.   10-3 Handhole with cover-plate.

Q What type of gasket is used on manhole and handhole cover plates?

A An asbestos gasket, molded to the correct shape, is commonly used. It should not be over $1/4$ in. thick when compressed.

Q How would you replace a manhole cover plate after washing out a boiler?

A It is not always necessary to replace the gasket, as a good one may be satisfactory after the joint has been broken several times. However, if there is any doubt that the gasket will make a tight joint, it should be discarded for a new gasket.

After cleaning off the joint faces, place the new gasket on the plate and smear the face of the gasket with a paste made from graphite and cylinder oil so that the joint will break readily the next time the boiler is opened. Insert the plate through the manhole opening. Hold it in place with one hand while positioning the yoke, and screw on the nut with the other hand.

Then put the second yoke on, if two are used, and tighten both nuts, making sure that the plate remains central in the manhole opening and that the gasket has not shifted. Large manhole covers in the ends of horizontal drums are sometimes supported on hinges inside the boiler head. These covers are not removed when the boiler is opened up but simply swung back clear of the opening.

Q Show, by sketches, the number and arrangement of washout plugs or handholes in a firebox water-leg type boiler.

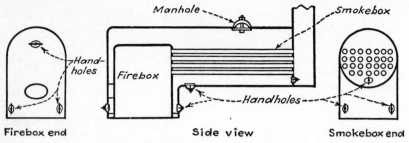

FIG. 10-4 Arrangement of handholes and washout plugs in firebox water-leg type boiler.

A The boiler should have not less than six handholes or washout plugs, located as in Fig. 10-4: one in the rear head below the tubes; one in the front head in line with the crown sheet; four in the lower part of the water leg. When possible, a seventh one should be near the throat sheet.

Handholes should be at least 2¾ by 3½ in. and washout plugs not less than 1-in. pipe size. Larger sizes are desirable, since the purpose of these openings is to permit visual inspection of the interior of the boiler and the insertion of washout hose and cleaning tools. Plugs of brass, bronze, or other nonferrous metal may be used where pressures are not over 250 psi. They can be removed more easily than steel plugs and are less likely to corrode or rust in place.

Q How are threaded connections attached to boiler shells?

A When the plate is fairly thick and the threaded opening is *very small,* the plate is not greatly weakened by the hole, so the pipe is simply screwed into the plate. If the opening is large and the plate too thin for a strong and leak proof threaded joint, a reinforcing pad may be riveted or welded to the shell and a continuous thread run through both pad and shell plate. Attachment of a blowoff pipe to an hrt boiler by this method is shown in Fig. 10-5. All such connections may be seal-welded either in-

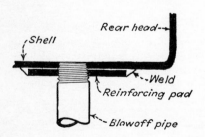

FIG. 10-5 Reinforcing pad for blowoff pipe.

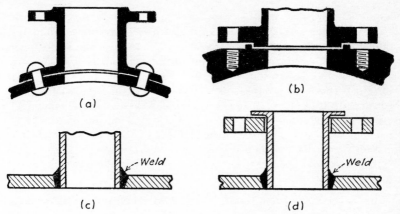

FIG. 10-6 Types of boiler nozzles used on steam boiler.

side or outside the shell, though this is not particularly desirable in the blowoff-pipe connection, which may have to be renewed frequently.

For boiler pressures over 100 psi, the maximum size of threaded pipe connections is 3 in.

Q Sketch several acceptable types of boiler-nozzle construction.
A Figure 10-6*a* shows a nozzle riveted to the shell; (*b*) shows a nozzle attached by studs screwed into the shell plate; (*c*) shows a plain nozzle, welded to the shell plate; and (*d*) shows a welded nozzle with loose flange.

INTERNAL INSPECTION

Q What defects would you look for when making an internal inspection of a steam boiler?
A After the boiler has been thoroughly cleaned, examine all plates, tubes, rivets, bolts, stays, and internal fittings for signs of corrosion, cracks, breakage, or distortion of shape. Cracked stays or rivets may be detected by tapping the parts lightly with a hand hammer, for there is a decided difference between the ringing sounds given off by a cracked part and by one that is not cracked. Corrosion is indicated by an eating away of the metal surface, and a great number of fine cracks may indicate *embrittlement*.

Q What defects would you look for when making an external inspection of a steam boiler?
A Such inspection should include the setting as well as the boiler proper, unless the boiler is self-contained. If the boiler has an external setting,

examine it for cracks or other signs of deterioration in the brickwork, particularly in the refractory lining of the furnace. Carefully examine the supporting columns, bolts, and beams to see whether they are sound. The outside of the shell and all attachments to it should also be examined closely for signs of corrosion, cracks, or breaks.

Carefully check parts subjected to intense heat for signs of burning, blistering, or bagging of plates because of weakness in the metal or overheating by accumulations of oil, scale, or mud on the water side of the plate. When any part of a boiler shell or attachment to the shell is concealed by a brick or other insulating covering, remove enough of this covering to examine plates, riveted joints, and points of attachment of fittings to the shell.

Q What is your duty when you discover any defects in the boiler or boilers under your charge?
A The answer depends largely on such local conditions as nature and size of plant, nature of load, plant personnel, system of boiler inspection, regulations regarding inspection of boilers, and distance from the boiler inspection headquarters.

If the defect is obviously dangerous, the engineer has no alternative. He must shut down the damaged boiler immediately and notify his employer and the boiler inspector. The inspector should then give instructions regarding the repairs or replacements necessary to put the boiler in safe working condition, and he should inspect and approve these changes before the boiler is again put in service.

Where the boiler inspector's services are immediately available, as is usual in thickly populated centers, the engineer should always follow the above procedure and leave matters to the inspector's judgment, although a shutdown may not actually be necessary unless the inspector orders it.

In sparsely settled communities or isolated plants far from any boiler-inspecting center, the stationary engineer, like the marine engineer, may sometimes have to rely upon his own judgment as to carrying on or making minor repairs before the inspector can be notified and inspection made. This is especially true where a shutdown may result in great inconvenience and perhaps human suffering; but under no circumstances should risks be taken that might result in a boiler explosion. Where it is absolutely necessary, in these isolated cases, to carry out minor repairs, notify some competent boiler-inspecting authority at the earliest possible moment.

Q What is *corrosion?* How is it caused and prevented?
A *Corrosion* is an eating away of the surface of plates, tubes, and rivets or stays by acid impurities in the boiler feedwater. Prevent it by changing over to a purer feedwater. If this is impossible, treat the feedwater so as

to neutralize the contained acids. Air dissolved in the feedwater is also a cause of corrosion, and a special apparatus called a *deaerator* is sometimes used to remove oxygen and other dissolved gas from the water before it enters the boiler. Some types of open feed-water heaters are also efficient deaerators, particularly those called *deaerating heaters.*

Zinc slabs, securely fastened to some part of the inner boiler surface, are sometimes used to reduce corrosion. The zinc instead of the boiler metal is attacked and eaten away by the galvanic action of the corrosive agents. When zinc plates are used, suspend metal trays beneath them to catch falling debris, or it may pile up on some part of the boiler heating surface, causing this surface to become overheated and burned.

Q What is *pitting?*
A This term is applied to corrosion that is not uniform over the entire metal surface, but occurs only in small spots or pits and gives the metal a honeycombed appearance. Pitting is caused by lack of uniformity in the metal. It may be that the parts which are eaten away contain impurities or material more susceptible to corrosion than the rest of the metal. Or the differences may set up local galvanic couples, as when copper and zinc are in contact in an acid solution.

Q What is *caustic embrittlement?*
A It is actual physical change in metal that causes it to become extremely brittle and filled with minute cracks, especially in the seams of riveted joints and around the rivet holes. Although hundreds of thousands of dollars have been spent in research, the exact causes of embrittlement are not fully understood. It is believed to be caused by the action of caustic soda in boiler water, either naturally present in the feedwater or introduced in excessive quantities by some form of chemical treatment. To prevent caustic embrittlement, feed-water causticity should be kept below the danger point by suitable chemical treatment, and all riveted joints calked on the inside.

CLEANING HEATING SURFACES _____

Q How is soot removed from the inside of fire tubes?
A Soot deposits are usually removed by hand scrapers made to fit closely in the tube or by steam blowers operated by hand or built into the boiler setting. A hand scraper is shown in Fig. 10-7. In the permanently installed soot-blowing attachment for fire-tube boiler in Fig. 10-8, the short vertical pipe carrying the blowing jets can be rotated through a semicircle by the outside handle. When the blower is not in use, the short horizontal pipe is slid back through the steamtight sleeve and the blower pipe placed

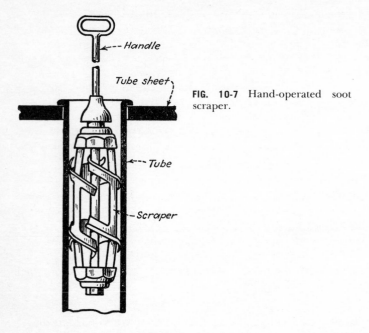

FIG. 10-7 Hand-operated soot scraper.

horizontally in the brickwork recess away from the direct sweep of flames and hot gas.

Blowers for water-tube boilers usually consist of horizontal pipes carrying a number of small steam jets set to blow downward vertically or diagonally between the tubes. The blower pipes are supported on top of the tubes and can be rotated from the outside of the setting.

Q How can boiler-scale deposits be removed by mechanical means?
A Remove soft-scale deposits on plates and tubes by hand scrapers, using special long-handled tools to reach parts not readily accessible. If the scale is hard it must be chipped off with chisels and hand hammers,

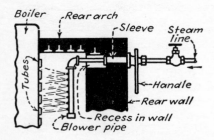

FIG. 10-8 Steam-jet soot blower installed in fire-tube boiler.

or with air hammers if compressed-air tools are available. Remove hard scale deposits on fire tubes by using an air- or steam-driven vibratory hammer. Water tubes are usually cleaned by compressed air, water, or steam-driven cutting tools.

Q Discuss the use of compressed air, steam, and water power for driving water-tube cleaners.

A The water-driven tube cleaner has a small water turbine of conventional design to rotate the cutter head, and the steam- and air-driven cleaners use a small rotary engine cylinder for this purpose. The form of cutting head and outside construction is similar in all three types.

The water turbine requires a large volume of water, and the high temperature of steam presents risk of injury to the operator, making its use inadvisable if other forms of power are available. Compressed air has none of the drawbacks associated with water or steam power, and its use is preferable to either, with water power as a second choice.

Figure 10-9 shows an air-driven water-tube cleaner fitted with a wheel cutter, and a cross-sectional view of the rotary motor used to drive the cutter head. This motor has only two moving parts, the shaft and a blade which fits into a slot in the shaft and can slide freely in the slot as the shaft revolves in the eccentric casing or cylinder. Compressed air enters through the inlet port, acts upon the sliding blade to cause the shaft to rotate, and then escapes through the exhaust port when it is uncovered by the sliding blade.

The cutter head shown in Fig. 10-9 has several arms carrying a number of hardened-steel cutter wheels. These arms swing outward by centrifugal force as the motor revolves the cutter head, and the sharp teeth of the cutter wheels cut away the scale. Other types of cutter head can be fitted to the motor for such operations as opening up a passage through a tube that is partially blocked by a thick accumulation of mud or scale before using the wheel cutter to complete the cleaning process.

Q Describe a power-driver cleaner for removing scale from fire tubes.

A The use of steam to drive fire-tube cleaners is objectionable because of

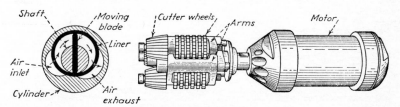

FIG. 10-9 Air-driven tube cleaner used inside of water tubes.

the risk of injury to the operator and because water is likely to damage the boiler setting; therefore compressed air is preferable for this purpose.

Figure 10-10 shows an air-driven cleaner in the process of scaling the outside of a fire tube. The hammer is connected by a swivel joint to the same type of rotary motor as that used in the water-tube air-driven cleaner. As the hammer head rotates, it strikes light and rapid blows all around the inner circumference of the tube, causing the outer coating of scale to crack and fall off. The cleaner should be kept constantly in motion so as not to concentrate the hammering action on one particular part of the tube for any length of time, as this might damage the tube, nor should it be operated too close to the tube sheet since it may tend to loosen the tube in the tube sheet.

Q Before internal inspections, how is the water side of hot boilers cooled down?

A Figure 10-11 shows methods of cooling down (*a*) fire-tube, (*b*) straight, and (*c*) bent-tube water-tube boilers. Entering hot boilers is a problem. It sometimes helps to change air flow so the heated stack draws fresh air through the boiler, but best result isn't always obtained with most accesses open. Assume that the fire-tube boiler must be entered 6 hr after cutout, as would be required in a creamery where milk is received 7 days a week and steam is needed until all equipment is washed—usually at 4 or 5 P.M. To get in, haul fire and open drafts to cool refractory; then haul ashes and wet down. While it is not good practice, pumping cold water in and blowing down may get steam off within an hour. Do this until you can hold your hand against the blowpipe. Drop in the top manhole cover and start spraying water inside the shell from the outside. Keep your head to one side to avoid the cloud of vapor from the manhole.

Refractories lose some heat after a few hours, but the boiler shell remains hot. Continue to spray and lower water in boiler. Drop front manhole cover, shown in Fig. 10-11*a*, and remove it. Now close front uptake doors. Some heat will be drawn from furnace and combustion chamber through tubes and go up stack. At the same time, fresh air is

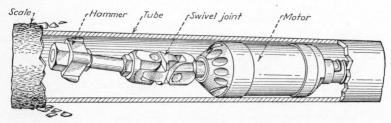

FIG. 10-10 Air-driven tube cleaner used for scale outside of fire tubes.

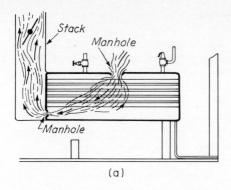

(a)

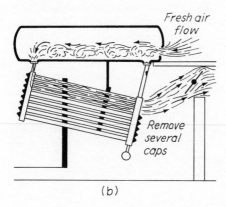

(b)

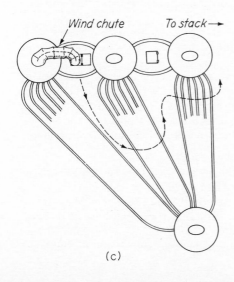

(c)

FIG. 10-11 Methods used to cool three types of boilers before entering waterside. (a) Fire-tube boiler; (b) straight boiler; (c) bent-tube boiler.

drawn into top manhole and goes down and out front manhole, picking up vapor on the way. Continue spraying. When vapor stops, spray for a few more minutes. The top of tubes under the manhole will dry quickly.

Longitudinal drum boilers, Fig. 10-11b, must have four or five hand-hole plates knocked out in rear header under the drum. Lay burlap over downcomer tubes in rear saddle. Close ashpit, cleanout, and firing doors. Circulation will be through manholes and length of drums, down through front nipples to front headers, back through tubes, and then through uptake or smoke trunk.

Bent-tube boilers are difficult to ventilate through the drums. You may make a wind chute of canvas and hoops, fit over a manhole. Place the other end at a pass cleanout door, Fig. 10-11c. Bent-tube boilers have to be well cooled before working inside.

While it is best to allow plenty of time to cool down boilers gradually, even boiler inspectors use these expedient methods when time is short.

SUMMER PROTECTION FOR HEATING BOILERS

Q Steam-heating boilers may deteriorate more during the summer shutdown than in an entire heating season unless properly prepared. List preventive maintenance at time of lay-up to retard corrosion and help assure trouble-free operation during the next heating season.

A Take these steps, where applicable, as soon as possible after the end of the heating season:

1. Remove all fuses from the burner circuit.
2. Remove soot and ash from the furnace, tubes, and flue surfaces.
3. Drain the boiler completely.
4. Flush the boiler to remove all sludge and loose scale particles.
5. Repair or replace leaking tubes, nipples, staybolts, packing, and insulation.
6. Clean and overhaul all boiler appurtenances such as safety valves, gage glasses, and firing equipment. Special attention should be given to low-water cutoffs and feed-water regulators to ascertain that float (or electrode) chambers and connections are free of deposits.
7. Check the condensate return system for tightness of components.

Q What precautions should be taken with steel boilers?

A Steel boilers should be left open and dry. A sign or tag must be placed on the unit to warn that it is empty and *not* be fired. Although dry lay-up is preferred for steel boilers (see "Standard Plant Operators' Manual" for detailed sketches, showing each operation), they may have to be kept ready for operation on short notice in some plants. The water should contain a suitable corrosion-inhibiting chemical and should fill the boiler completely. To be sure, consult your boiler-water-treatment specialist.

Q What precautions do cast-iron boilers need?

A Cast-iron (c-i) sectional boilers should be filled to the top with water. A sign or tag should be attached cautioning that the boiler must be drained to normal water level before firing. Internal corrosion in c-i boilers may not be a serious problem. But if the water used is highly corrosive, follow lay-up procedure for steel boilers.

If the c-i boiler is exposed to humid atmospheres and rapid temperature fluctuations, dry lay-up is best to prevent sweating and thus corrosion of external surfaces.

Q How should the boiler be returned to service after the summer lay-up season?

A Fill with water to normal operating level. Replace fuses. Immediately after firing the boiler, test all automatic controls including feed-water regulator, low-water fuel cutoffs and alarm, safety valves, and combustion safeguards.

> CAUTION:
> *Do not* leave an automatically fired boiler unattended after the initial start-up until it has gone through several firing cycles and you are sure that all controls are functioning properly.

Q How should hot-water-heating boilers be protected for the summer season?

A Most of the steps recommended for laying up steam-heating boilers also apply to hot-water-heating boilers. An exception is that complete draining and flushing are usually *not* necessary. The recommended procedure is to drain from the bottom of the boiler while it is still hot (180 to 200°F) until the water runs clean, then refill to normal water-fill pressure. If water treatment is used in the system, sufficient treatment compound should be added to condition the added water.

SUGGESTED READING

REPRINTS SOLD BY *Power* MAGAZINE
  Steam Generation, 48 pp.

BOOKS
  Elonka, Stephen M., and Anthony L. Kohan: "Standard Boiler Operators' Questions and Answers," McGraw-Hill Book Company, New York, 1969.
  Elonka, Stephen M.: "Standard Plant Operators' Manual," McGraw-Hill Book Company, New York, 1975.
  Elonka, Stephen M., and Joseph F. Robinson; "Standard Plant Operators' Questions and Answers," vols. 1 and 2, McGraw-Hill Book Company, New York, 1959.

11

BOILER REPAIRS, FOUNDATIONS, AND ERECTION

While stationary engineers are seldom called upon to erect boiler and machinery foundations or furnace brick walls, operators and boiler owners in many remote areas often must perform these jobs themselves. Examples are mines and other industries using boilers in remote areas, or developing countries where experienced help is far away. Repairs must be made on boilers in similar circumstances.

The boiler operator and stationary engineer are often the best qualified persons in such situations, much like the marine engineer who must make unusual repairs in emergencies at sea in order to bring the ship safely into port.

BOILER REPAIRS

Q What kind of high-pressure boiler repairs may be made by electric or oxyacetylene welding?

A Factory welds can be tested and flaws readily detected by x-ray and other forms of testing apparatus. These testing methods cannot be applied easily to the common run of plant repair jobs, so exercise great care and judgment when carrying out welding repairs on high-pressure boilers. In general, it is inadvisable to weld long cracks in plates that are exposed to intense heat or to high tensile stress, but in many other cases risk of failure is not so great, and welding is more convenient than other forms of repair. Because much more depends upon the human factor in welding than in other classes of mechanical repair work, it should be done only by skilled, experienced welders.

Some common defects that may be welded satisfactorily are: short cracks in internally fired furnaces; cracks in stayed surfaces, such as side sheets of water legs; cracks in tube sheets; cracks in headers.

Always ask the advice of boiler-inspection authorities before attempting any welding repairs, and follow their instructions closely.

Q When a damaged plate is to be repaired by welding in a new piece, how are the plate edges and patch piece prepared for welding?

A When the patch is in place, plate edges and patch piece must be beveled to form a vee which is then filled in with the welding material. If any stays or rivets pass through the part that is cut away, they must be inserted in the same position in the patch plate.

Q How is a repair made if part of the bottom of the shell of a riveted fire-tube boiler has to be cut out and a patch put in?

A If the shell bottom is directly exposed to intense furnace heat, and welding would probably not be allowed, the patch has to be riveted. Cut out the damaged part to leave an elliptical or diamond-shaped hole with rounded corners. The patch should be cut to the same shape with sufficient overlap to allow for riveting, bent to the same curvature as the boiler shell, fitted in place inside the shell, and riveted to it by rivets of the same diameter as those in the shell joints. The rivet patch should be proportioned to give the patch joint strength equal to or greater than the other shell joints. Edges of plate and patch should be calked.

Q How are cracked ligaments between tubes repaired?

A The old method was to make a patch plate called a *spectacle piece* because of its shape, and pin or rivet this to the tube plate over the crack, but now welding does a much better and neater job. Tubes on either side of the crack are removed, the crack vee'd out by chipping or grinding, and then welded in the usual way. After welding is completed, edges of the weld in the tube holes and the surface of the tube sheet are smoothed by chipping and filing, or grinding, and the tubes replaced.

Q Name the advantages of preheating and annealing work that is being, or has been, welded?

A Advantages of preheating are obvious, especially when large parts are welded. If the metal is preheated to a fairly high temperature, heat from the torch is almost entirely available for melting the filler rod and the metal at the edges of the joint. Preheating also causes the metal to "give" and thus avoids expansion and contraction stresses at the weld or in the surrounding metal. If the metal is cold, on the other hand, heat from the torch is rapidly conducted away by the cold metal, and it is more difficult to keep the metal in the joint at a welding heat and do a satisfactory job. Rapid cooling also tends to make the weld metal extremely hard and difficult to chip, file, or machine.

Annealing a part after it has been welded allows the strains that have been set up by unequal heating to equalize and adjust themselves. There is also less danger of an annealed part breaking at the weld or close to it when again put into use. Annealing is done preferably in a proper annealing furnace where temperature is under close control. If such a furnace is not available, the part may be heated to a dark red and then buried in sand, lime, or ashes so that it cools very slowly.

Q How do you remove a defective tube in a fire-tube boiler if the tube is in the center of the tube sheet?

A Since the tube is surrounded by other tubes, it must be pulled out through the front tube sheet. The beads are cut off both tube ends and the ends split in three or four places by means of a thin round-nosed chisel or a diamond point. Take great care not to cut or otherwise damage the tube sheet. The split tube ends are then closed in a little with a blunt-nosed tool to loosen them in the tube sheets, and the tube is driven out of the tube sheet, using a piece of pipe or bar of suitable diameter. If the tube is covered with scale, it may be necessary to use a set of blocks to pull it out through the front tube sheet. Figure 11-1 shows the method of ripping the tube end with a diamond-point chisel after the bead is cut off.

Q How do you remove a defective tube in the bottom row of a fire-tube boiler?

A You can avoid the laborious job of pulling the tube out through the front tube sheet by cutting the tube off just inside of both front and rear tube sheets, dropping it down, and pulling it out through the front manhole or handhole. The ends left in the tube holes are then split and knocked out. This is the fastest and easiest way to remove tubes when a boiler is being retubed. An internal wheel cutter which resembles a tube expander in construction and operation may be used to cut off the tube ends or, if this is not available, a simple hook chisel bar. The method of using this bar is shown in Fig. 11-2.

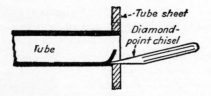

FIG. 11-1 Ripping end of a tube.

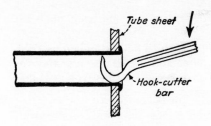

FIG. 11-2 Using a hook- cutter bar.

Q Explain the correct procedure for putting a new tube or set of tubes in a fire-tube boiler.

A After the old tube or tubes are removed, the holes in the tube sheet are cleaned and any ragged or sharp edges rounded off slightly. The tube ends are annealed to make the metal ductile and prevent cracking when the ends are being expanded and beaded. In an hrt boiler, the holes in the front tube sheet will be an easy fit for the tubes, and the holes in the rear tube sheet will be a driving fit unless the holes have been enlarged by excessive tube expanding.

After the holes are properly prepared, the tube ends are cleaned and the tubes driven into place until the ends project about $1/4$ in. beyond the tight tube sheet. The tight ends are now expanded and beaded. Best procedure is to tighten the tube in the tube sheet with the tube expander just enough to keep it from moving. Next splay the tube end out to an angle of 45° with a ball-peen hammer, Fig. 11-3a, bead it back on the tube sheet with a beading tool, Fig. 11-3b, and finally, finish expanding the tube. If it is fully expanded before beading, it will probably be loosened up a little by the beading process and require retightening.

When all the tubes are expanded and beaded in the tight tube sheet, the opposite ends are cut, if necessary, to the proper length for beading, then expanded in the same way. Sometimes beading is dispensed with if the tube ends are not directly exposed to intense heat, as in the smokebox tube sheet of a fire-tube boiler. The main object of beading is to prevent

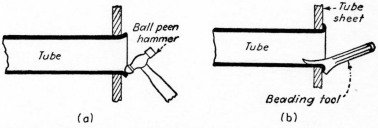

FIG. 11-3 Splaying end (a) and beading end (b) of boiler tube.

the tube end from burning off. It is the expanding that really holds the tube in the sheet.

The welding of beads to the tube sheet is also a fairly common practice. This cuts down leakage at the tube ends and avoids the necessity for frequent reexpanding which is often characteristic of this type of boiler.

Q Describe some type of tube expander for boiler tubes.

A The self-feeding roller tube expander is most commonly used for this purpose. The body of the roller tube expander, or cage, fits loosely in the tube, and the expanding rolls are pressed outward by a tapered mandrel. When expanding is done by hand, the mandrel is turned by means of a short bar inserted through a hole in the large end of the mandrel. When mechanical power is used, the end of the mandrel is shaped to fit into a socket in the driving tool. The expanding rolls are held loosely in the cage and are free to turn as the mandrel revolves. They are set at a slight angle to the longitudinal axis of the expander, and this tends to feed both the mandrel and the cage inward.

The expander in Fig. 11-4 is power-driven and has a set of flaring rolls in addition to the expanding rolls. These flaring rolls flare out the part of the tube that projects beyond the edge of the tube plate, a common practice where tube ends are not exposed to fire and beading is therefore not necessary. This expander also has a stop collar threaded on the cage extension. This collar can be adjusted to give the expander just the correct amount of inward travel to expand and flare the tube end fully. When the collar comes in contact with the tube sheet or header, the tube should be fully expanded and a few more turns is all that is needed to set the tube and complete the operation. In simpler types of tube expanders with no stop-collar adjustment, the operator has to depend upon the feel of the expander to know when expansion is complete, and this feel can only be acquired by practice. In order to do good work, the expander must be kept well lubricated and both expander and tube end must be clean and free from dirt or scale.

Q What should you do when a new tube fits very loosely in the holes in the tube sheets?

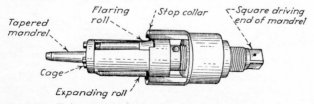

FIG. 11-4 Tube expander of the self-feeding roller type.

A In an old boiler where leaky tubes have perhaps been expanded over and over again, the holes in the tube sheets may become enlarged to such an extent that a new tube will be a very loose fit. Here the new tube should not be expanded out until it tightens in the tube sheet, as this stretches the tube too much and weakens it so that it will not remain tight.

Fill the space between the tube and the tube sheet with a thin metal bushing called an *outside ferrule*. Ready-made seamless copper ferrules can be bought for this purpose. If they are not easily procurable, a thin strip of copper plate, flattened to a taper at the ends and bent in a circular shape so that the tapered ends overlap, does very well. The ferrule is driven into place between tube and sides of tube hole, and the tube is then expanded and beaded in the usual way. Inside ferrules are sometimes driven into the ends of old leaky tubes as a temporary means of stopping leakage. However, they reduce the tube area and interfere with cleaning.

Q How do you remove and replace a defective stay bolt?

A Cut off the bolt heads. Drill a hole in each end slightly deeper than the thickness of the plate, using a drill that leaves only a thin shell of metal in the sheet. Close in this metal shell with a round-nosed chisel, taking care not to damage the threads in the plate. The piece of stay bolt in the boiler then drops down into the water leg, and can be fished out through a handhole.

If the threads have not been damaged, put in a stay bolt of the same size as the old one, but if the threads are spoiled it may be necessary to run an oversized tap through the holes and put in a larger-diameter stay bolt. In either case, the new bolt is screwed in, cut off on each end, leaving a few threads for riveting, and riveted on each end to form a head. When one end is being riveted, a heavy hammer is held against the other end. It is important that the new bolt be a tight fit in the threads. If it is loose to begin with, it will leak in a short time and the threads may possibly corrode in the plate.

A skillful welder can burn out the center of an old stay bolt with a torch in much less time than it takes to drill it out, but he must exercise great care not to burn or melt the plate. Drilling, though slower, is probably safer.

Q How is a hand ratchet set up for drilling holes?

A If there is a convenient place for fastening it to a bolt or stud, a boring post is set up, as in Fig. 11-5. The base is bolted rigidly to the job and the adjustable arm clamped at the right height for the ratchet. The drill is fed into the work by the feed screw on the opposite end of the ratchet. If a boring post cannot be set up in this fashion, place some heavy object

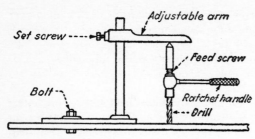

FIG. 11-5 Setting up hand ratchet for drilling.

against the ratchet to hold it, or set up and brace a plank from an adjacent wall or another boiler or machine. To apply pressure to the drill, use the same methods as for electric or air-driven drills.

Q How are water tubes fastened in tube sheets and headers?
A Water tubes are secured the same way as fire tubes by expanding the ends with a tube expander, but the projecting ends are usually left untouched or simply splayed out a little without being beaded except when tubes enter the bottom of a drum. Ends of these tubes may be beaded to offer as little obstruction as possible to flow of water into the tubes, and to drain the drum completely when the boiler is emptied for cleaning, inspection, or repair.

Q How are defective water tubes removed and replaced?
A Ends are split with a narrow chisel, then bent inward with a crimping tool, which is a short bar with a narrow slot in the end that fits over the tube. Straight tubes may then be pulled out through one of the tube holes. This is not always possible with bent tubes, whose ends may have to be worked out of each tube hole in turn and the body of the tube removed sideways. If the defective tube in a bent-tube boiler is in an inner row, it may be necessary to remove several tubes to get at the damaged one. This may require replacing almost an entire row instead of only one tube, unless the spacing permits removal of a single tube.

FOUNDATIONS

Q Explain laying out a foundation for a battery of boilers.
A First outline the foundation wall on the ground with reference to some base line, such as the boiler-room wall. Use a measuring tape, preferably steel, and a large square. Corner markers, which are set well back from the excavation edge so that digging the trench won't disturb them, may be wooden stakes. However, *batter boards,* Fig. 11-6*a,* are

better, and their tops should be leveled. Then lines are run to guide the excavation work and provide checking points for height of foundation walls, Fig. 11-6*b*.

Depth and width of excavation depend on the soil. In solid ground, a shallow trench may be all that is necessary to reach *hardpan.* If really solid footing means going to a great depth, and the ground is fairly firm and well drained, it may be possible to ensure a stable foundation by widening the trench and placing a concrete slab on the bottom. This footing slab, or *mat,* should be at least 18 in. thick and reinforced with steel or iron bars.

When the ground is soft to a considerable depth, and a more suitable location cannot possibly be selected, you may have to drive a large number of piles and then run a wide reinforced-concrete mat over their tops to give a still greater bearing area.

Q What are average bearing capacities, in tons per square foot, of rock, gravel, sand, and clay?

A No exact figures can be given because these materials vary greatly in composition, and their bearing capacity is strongly affected by presence or absence of moisture. Various competent authorities give these average values:

Nature of ground	*Safe bearing strength, tons/sq ft*
Hard rock	40
Soft rock	8
Hard dry clay	4
Wet clay	1
Cemented gravel (hardpan)	10
Gravel	6
Coarse dry sand	4
Wet sand	2

Q How do you make and set forms for a concrete foundation?

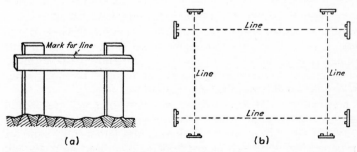

(a)   (b)

FIG.   **11-6** Boiler foundation layout.

A Make them of common boards, at least ⁷/₈ in. thick, with one or both sides smooth. Nail these boards, smooth side inward, to two-by-fours, making the joints so tight that concrete can't leak through. Place forms in the trenches on the footing slab, so braced to each other and to the sides of the trench that they cannot move when the concrete is poured. The space is filled with concrete, and the top is smoothed off to provide a level footing for the brick walls.

Bottom bolts for columns are embedded in the concrete if steelwork is to support the boilers, and the whole mass is allowed to set a few days before the forms are removed and the walls built. Make sure concrete is set hard before removing the forms.

Carefully check alignment of the forms with the base line and dimensions on the plan before pouring.

Q What are proportions of sand, gravel, and cement for boiler or engine foundation?

A Strength of a concrete mixture depends on nature and proportions of ingredients and on mixing method. Proportions of cement, sand, and gravel again depend upon size and type of machinery to be supported. Less cement is needed if the aggregate (crushed rock or gravel) is of uniform size.

The strongest concrete mixture commonly used runs about $1:1\frac{1}{2}:3$, indicating proportions by volume of cement, sand, and gravel in that order. The weakest mixture runs about $1:3:6$. The first mixture would be suitable for foundations for heavy vibrating machinery, the second for light smooth-running machines. Boiler foundations carry heavy weights, but are not often subjected to excessive vibration, so a medium mixture, say $1:2:4$, should usually be satisfactory.

For maximum strength, use clean sand of good quality and aggregate composed of gravel or crushed hard rock. Cinders, soft rock (like limestone), or dirty clayey sand produce a weak concrete and may ruin the foundation, causing great trouble and expense later.

Q How do you mix concrete for a boiler or engine foundation?

A If a mechanical mixer is available, the proper amounts of sand, gravel, and cement for one batch are thoroughly mixed while dry. Then only enough water is added to make the concrete sufficiently plastic to fill all spaces when it is poured and tamped into the forms. Too much water tends to float out cement and weaken concrete.

For hand-mixing, dump the measured amounts of sand, gravel, and cement on a large watertight board platform and mix thoroughly while dry. Then add enough water to make the mixture into a stiff plastic for further mixing. Main essentials of good concrete are thorough dry and wet mixing, clean sand and gravel, and not too much water.

Q What proportion does the dry material bear to the mixed concrete?
A Separate dry ingredients of concrete always occupy more space than the wet mixture. Shrinkage averages 25 percent. The right amount of water in mixing is important.

Typical proportions of cement, sand, gravel (or crushed rock), and water for a 1:2:3 mixture are

6 sacks cement (approx 6 cu ft)
12 cu ft sand
18 cu ft gravel or crushed rock
38 gal water (6⅓ gal per sack of cement)

Q How do you pour concrete in boiler foundation walls?
A It is common practice in a small job to load the concrete into wheelbarrows by hand shovel, or direct from the mixer if a machine is used, and to arrange plank tracks so that the concrete can be wheeled and dumped directly into the forms. As the concrete is poured, it is tamped with an iron bar to make sure no holes or cavities are left unfilled. This is particularly necessary when the concrete is mixed fairly stiff.

Do not run concrete in freezing weather if it can possibly be avoided. If it must be run in cold weather, heat the sand and gravel, use warm water in mixing, and cover the concrete with some heavy material to prevent freezing before it starts to set.

When the forms are filled to the required height, place the column foundation bolts in position, if supporting steelwork is to be used. Then smooth the top of the concrete to provide a level base for the brickwork. Allow the concrete to set properly before the forms are removed and the walls built.

Q How do you erect boilers after foundation walls are finished?
A When the concrete in the walls has set hard, the forms are removed and the excavation filled in. The ashpit and combustion chamber floor can now be laid and allowed to set. The next step depends on the types of boilers being erected. Haul hrt boilers into position over the foundation, and jack up and support on blocks about where they will finally stand. Small hrt boilers are usually supported by brackets resting on the side walls. Suspend large hrt boilers from steel beams supported on steel columns.

If the boilers are to be supported by brackets resting on the side walls, walls are built up to the finished height and the blocking removed. If steelwork is used, the supporting columns and crossbeams are placed in position and bolted to each other and to the foundation bolts. The suspending bolts can also be attached to the crossbeams and the boiler hangers; but do not remove the blocking underneath the shell until the columns are braced firmly in every direction.

Carry brickwork as high as possible before removing the braces and blocking, and take great care when removing the blocking to prevent any rolling or swinging of the shell that would throw a sudden load or side thrust on columns and braces. When the bottom blocking has been removed, carry the walls up to full height and finally discard the column bracing.

Practically all multidrum water-tube boilers are supported by steelwork erected before the drums are set in place and the setting built. Observe the same precautions in securely bracing columns and beams, so there is no possibility of movement during the assembling of boiler parts and the building of the enclosing setting.

Q Describe the solid-wall boiler setting.

A This is probably the commonest form of setting wall construction in small and medium-sized boiler plants. Interior of the combustion chamber and furnace is lined with first-grade firebrick or a plastic refractory material made especially for this purpose. Outer wall is constructed of second-grade firebrick or a good quality of common red brick. The firebricks forming the furnace lining are not laid in mortar, but are merely dipped into a thin mixture of fire clay and water and rubbed into place so as to leave as thin a joint as possible. Common practice is to make the firebrick lining one brick thick, with a row of headers (bricks laid endwise) every fifth or sixth row to bond the lining to the outer wall.

Bricks in the outer part of the wall are usually laid in a mortar of lime and sand, or of cement, lime, and sand in the approximate proportions 1 : 1 : 5 by volume. Thickness of the setting walls depends on size of boiler and weight (if any) to be carried by the walls.

Q How are plastic materials used in the construction of boiler settings?

A Plastic refractory materials for lining boiler settings are usually supplied by the manufacturer in the form of moist slabs packed in airtight cartons or containers to prevent drying out before use. Metal anchors of various shapes are used to tie the plastic material to the outer wall, and the lining is built by placing the plastic slabs against the furnace wall and pounding each successive row firmly into place with a heavy hand hammer or air-driven ram. When finished, the lining is trimmed off so as to present a smooth surface. Being moist and soft to begin with, the plastic material forms a homogeneous, monolithic mass with no cracks or joints. After installation, a plastic lining should be allowed to set for a few days, and then should be baked out with a slow fire to complete the hardening process.

Plastic refractories have several advantages over solid brick or tile. There are no joints to offer an opening for furnace heat to crumble or damage the lining. Repairs can be made to damaged plastic or brick lining without having to renew the entire lining. Odd shapes such as arches and baffle walls that would require special brick or tile and would be difficult to build are easily molded in place from plastic refractories.

Refractory material of similar composition to the plastic slabs can be procured in dry form. It is mixed with water to any required consistency and used in places where the material has to be cast or poured like concrete.

Q What are buck stays?

A Used to strengthen and support walls of brick boiler settings, buck stays are long, narrow steel or iron castings, flat on one side with a heavy

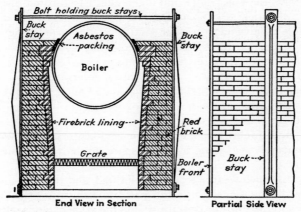

FIG. 11-7 Solid wall boiler setting with buck stays.

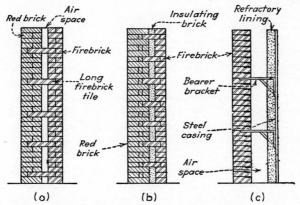

FIG. 11-8 Three types of furnace-wall construction.

stiffening rib on the other. They are placed with the flat side against the walls on both sides of the setting and held together by long bolts passing under and across the setting. Two or more pairs of stays support the side walls. Sometimes stays are also placed against the rear walls, with bolts extending the entire length of the setting to the boiler-front plate castings. Figure 11-7 shows side stays.

Q What are the disadvantages of the solid-wall boiler setting, and what other forms of construction overcome them?
A The solid-wall setting has two serious disadvantages: the heat loss through the wall is considerable, and the extreme difference in temperature between the wall's outer and inner sides has a strong disintegrating effect on the brickwork. Some wall constructions, Fig. 11-8, designed to offset these disadvantages without sacrificing too much strength are:

a. Brick wall having a firebrick lining and an air space between the lining and an outer red brick or second-grade firebrick wall, with long bricks or tiles tying the two walls together at intervals.

b. Brick wall having a firebrick lining and an outer wall of red brick or second-grade firebrick, with a layer of some kind of special insulating brick between lining and outer wall.

c. An inner firebrick wall carried on steel bearers, separated from an outer insulated steel casing by an air space through which air circulates constantly.

Wall (a) is much weaker than the solid wall, and the reduction in heat loss is hardly great enough to justify its use. Construction (b) cuts down heat loss and does not weaken the wall to the same extent as the dead air space. The air-cooled setting in (c) is a more efficient construction than either of the first two. It also serves as a preheater for combustion air.

Q What are super refractories?
A Super refractories are used in the fuel-bed area of furnace walls. They are an 85 percent silicon-carbide brick that has excellent abrasion resistance. Being very dense and not subject to corrosion by most coal ash, they resist clinker adhesion and erosion far better than fireclay.

For coals with ashes that fuse easily, silicon carbide works best when placed to fuel-bed depth because it resists abuse and wear. In higher-temperature regions above the fuel bed, harmful chemical conditions caused by floating ash particles containing much iron may cause slagging of silicon-carbide bricks. This doesn't happen in the cooler zone lower down.

For upper walls, the best refractories are made of mullite, which is a composition of aluminum and silicon oxides. This refractory is ideal because it forms inert crystals of mullite that prevent fluxing with ash. To ensure the correct composition, mullite is produced artificially from electric-furnace melts.

Bonded mullite and bonded silicon-carbide brick both stand temperatures well above 3000°F. Mullite has abrasion resistance much greater than firebrick. Its resistance to clinker attack exceeds that of silicon carbide a little, but firebrick by a wide margin.

Some shapes are laid, hung, wedged, or bolted in place. They should be laid in cements of their own composition with thin joints and carefully bonded into the firebrick backing. In most cases, they are installed as standard brick shapes and laid with standard bricklaying methods. When chosen and applied correctly, their life is several times that of common refractories.

WATERWALLS

Q What are waterwalls?

A They are rows of vertical or inclined water tubes lining the inner walls of boiler furnaces, and connected top and bottom to the other parts of the boiler so that there is continuous rapid circulation of water through the waterwalls. They provide a large additional area of boiler heating surface and also protect the refractory from the intense furnace heat. Waterwall tubes may be:

1. Plain tubes set close together to present a metal wall almost solid, or spaced a few inches apart, and backed in both cases by a refractory wall. Figure 11-9 shows the latter construction.

2. Tubes having metal fins welded to the sides and installed so that the fins touch and thus expose a complete metal surface to the furnace heat.

3. Tubes that have a great number of short metal studs welded on their sides and that are backed by and covered with a refractory material so that only the stud ends are exposed to the heat.

In stoker-fired boilers, waterwall tubes at the sides of the grates, and for a few feet above, are often protected by cast-iron blocks bolted to the tubes.

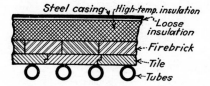

FIG. 11-9 Waterwall cross-section.

Q How are hrt boilers supported?

A Up to and including 54 in. diameter and 14 ft length, hrt boilers may be suspended from steel crossbeams resting on steel columns set outside the setting walls, or may be supported by four steel or cast-iron brackets riveted or welded to the shell and resting on the side walls of the setting. Boilers from 54 to 72 in. diameter may be supported by steel columns, beams, and hangers or by eight steel or cast-iron brackets set in pairs, four on each side and resting on the side walls. Boilers over 72 in. must be supported by the outside suspension method.

With bracket supports, front brackets rest on steel plates set on top of the brickwork, but rollers are placed between rear brackets and plates to confine movement due to expansion and contraction to rear end of boiler.

The bracket support is shown in Fig. 11-10 and the outside suspension method in Fig. 11-11.

Q Sketch in detail one method of attaching suspension bolts to the boiler shell.

A Figure 11.12 shows a suspension bolt attached to a steel lug welded to the shell.

Q Sketch the method of attaching suspension bolts to supporting cross-beams.

A Figure 11-13 shows a detailed sketch.

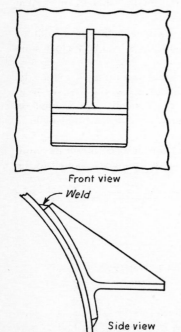

Front view

Weld

Side view

FIG. 11-10 Supporting bracket used on older type hrt boilers.

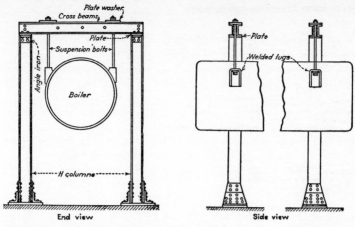

FIG. 11-11 Supporting steelwork of hrt boiler supports weight of water-filled drum above fire.

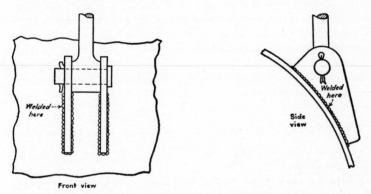

FIG. 11-12 Supporting lug is welded to the shell of boiler.

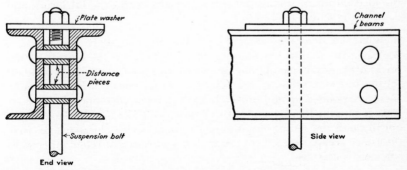

FIG. 11-13 Method of attaching suspension bolts to crossbeams.

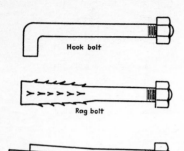

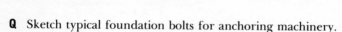

FIG. 11-14 Foundation bolt types.

Q Sketch typical foundation bolts for anchoring machinery.
A Three foundation bolts are shown in Fig. 11-14. The hook bolt is easily made but does not have the gripping power of the rag or wedge bolts. Hook and rag bolts are commonly used where the bolts can be placed in position before the concrete is run. Use the wedge bolt when the holes for foundation bolts have to be drilled in solid concrete. When a wedge bolt is placed in a drilled hole, drive it down on the wedge, taking care not to damage the bolt threads. This expands the split head so that it grips the sides of the hole.

FIRING NEW BOILERS

Q When putting a new boiler into service what precautions must be observed with the setting?
A Start a light fire after the boiler has been filled with water. The setting should be allowed to dry out under this fire for several days. This gives the walls a chance to settle and lets any strains equalize. Firing up with a full load on a new setting will almost certainly crack the brickwork and damage the setting badly.

Q What care does a boiler setting need during operation?
A When the setting has no outer steel casing, give the outside of the brickwork a heavy coat of paint or cover with a plastic sealing preparation for this purpose, keeping a watchful eye out for cracks or breaks. Cracks should be pointed up and broken brickwork repaired at once, for cold-air leakage into the furnace lowers its efficiency considerably. Leakage of gas out through the setting into the boiler room is also objectionable. Give the same care to the inside of the setting, and repair or reline inner walls whenever necessary.

SUGGESTED READING _____

REPRINTS SOLD BY *Power* MAGAZINE
Steam Generation, 48 pp.

BOOKS
    Elonka, Stephen M.: "Standard Plant Operators' Manual," McGraw-Hill Book Company, New York, 1975.
    Elonka, Stephen M., and Anthony L. Kohan; "Standard Boiler Operators' Questions and Answers," McGraw-Hill Book Company, New York, 1969.

12

FUEL AND HOW IT BURNS

The reaction by which fuel and oxygen combine and give off heat is still the single most important energy-producing process in the world. Heat is the operating engineer's stock in trade, and combustion is the process the engineer lives with daily. As fuels get more expensive, and as industry demands more power and more steam, the stationary engineer faces the need to understand better how fuels burn and how to get the most out of them without polluting the atmosphere. This chapter gives the basics on combustion chemistry that all persons working with fuel should know.

CHEMISTRY OF MATTER

Q What is *matter*?
A *Matter* is the general name for all the material substances, gaseous, liquid, or solid, forming the earth and its surrounding atmosphere.

Q What is the composition of matter?
A All matter is composed of simple substances called *elements*, or combinations of these elements. For example, by combining the two simple substances iron and carbon we get steel. Water is a combination of two gases, hydrogen and oxygen. There are 92 elements, ranging from hydrogen, the lightest element, to uranium, the heaviest.

Elements seldom occur in their pure state but are usually combined with other elements in varying proportions to form the infinite variety of material things in our universe.

Q What is an *atom?*

A An *atom* is the smallest particle of matter that can take part in a chemical change. It is composed of yet smaller particles called *electrons,* and the number of electrons in the atom determines its weight. Hydrogen, the lightest element, has one electron rotating around a *proton,* whereas uranium has 92 rotating electrons.

Q What is meant by *atomic weight?*

A This term refers to the *comparative* weight of the atom. For convenience, oxygen is usually taken as 16 on the atomic-weight scale, and the weights of the other atoms are given as compared with oxygen. This makes the atomic weight of hydrogen 1.008, or slightly more than 1.

Q What is a molecule?

A A molecule is the smallest particle of matter that can exist alone. It is composed of two or more atoms of the same or different elements. For example, a molecule of oxygen is composed of two atoms of oxygen, and a molecule of carbon dioxide is composed of one atom of carbon and two of oxygen.

Q What is meant by molecular weight?

A Molecular weight, or weight of a molecule, is calculated by adding the atomic weights of the atoms contained. Thus, the atomic weight of oxygen is 16, and the oxygen molecule contains two atoms, so the molecular weight of oxygen is $16 \times 2 = 32$.

Q What is a chemical combination?

A It is the combination of atoms of two or more different elements to form another substance, very often having entirely different physical properties. For example, two atoms of hydrogen gas combine with one atom of oxygen gas to form one molecule of water.

In a given chemical compound the atoms always combine in the same proportions. Their combining is a chemical change, and a chemical process is required to break up the combination into its component atoms.

Q What is a mechanical mixture?

A A mechanical mixture is a physical mixing together of two substances and does not entail any chemical change. A mechanical mixture may be again split up into its component parts by some mechanical or physical process, such as screening or washing. For example, we could mix sugar and salt and then separate them by washing out the sugar. This would be a purely mechanical process.

Q How do you usually denote the composition of a complex substance?
A Writing out the names of the elements in full every time they are used would be rather a laborious process, so we use a system of symbols instead. Usually the symbol is the first letter, or letters, of the name, or its Latin equivalent. Thus, for carbon we write C, for oxygen O, and for iron Fe, from the Latin word *ferrum.* If more than one atom of an element is to be denoted, a small figure, called a *subscript,* is placed after and below the letter; thus O_2 means two atoms of oxygen, O_3 means three atoms, and so on.

Any combination of elements can be shown very simply and plainly by these symbols and subscripts. For example, hydrogen sulfide is a chemical combination of hydrogen and sulfur, consisting of two atoms of hydrogen and one atom of sulfur in each molecule. All of this information can be conveyed briefly by writing H_2S. If more than one molecule is to be denoted, a large figure is placed in front of the symbol. Thus, $2H_2S$ means two molecules of hydrogen sulfide.

Q Name the most common elements in fuels and give their chemical symbols and atomic weights.
A

Element	Chemical symbol	Atomic weight
Hydrogen	H	1.008
Carbon	C	12.005
Nitrogen	N	14.01
Oxygen	O	16.00
Sulfur	S	32.06

These are the atomic weights, taking oxygen as 16. In calculations where atomic weights are used, the fractions are often neglected, as they are too small to make any practical difference in the answers.

Q What is air?
A By volume, the air we breathe is 21 percent oxygen and 79 percent nitrogen. The nitrogen is inactive, slow to react chemically with other substances. Not so the oxygen. The oxygen of the air is always busy tarnishing silver, coating copper with green, rusting iron and steel. All these processes are oxidation, which is the combining of substances with oxygen. All give off heat, but too slowly to start a fire.

COMBUSTION

Q What is combustion?

A Combustion is high-speed oxidation — so speedy that the heat of reaction keeps relighting the unburned part of the fuel, keeping the flame or burning continuous. Our eyes seem to tell us that wood, coal, and gasoline burn. Yet, strictly speaking, nothing truly burns unless it is a gas. When we burn coal, wood, or gasoline, we really burn gas produced from those solids or liquids.

A candle proves the point. The solid wax won't burn, Fig. 12-1. Even the liquid wax in the wick won't burn. But the vapor formed from this liquid as it rises up the wick to meet the flame does burn and generates further heat to melt and vaporize the next layer of wax. So the candle keeps burning as the wick sucks up new liquid to be vaporized and ignited by the heat of what went before.

Q How does gas burn?

A The gas must be combustible. It must be mixed with air in proper proportions. The mixture must be raised to the ignition or firing temperature and held there.

The simplest combustible gas is the element hydrogen. Chemists tell us that two atoms of hydrogen combine with one atom of oxygen to produce H_2O, which is ordinary water. When generated by combustion at high temperature, this water is first an invisible vapor that may later condense as liquid water. For every pound of hydrogen burned to water, 62,000 Btu of heat energy is given off.

Burning 1 lb of hydrogen uses up 8 lb of oxygen to produce 9 lb of water. The large amount of nitrogen carried along with the oxygen takes a free ride and does not enter into the chemical reaction.

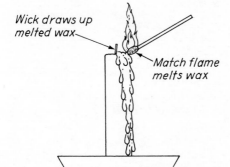

Wick draws up melted wax

Match flame melts wax

FIG. 12-1 Wick draws up melted wax by capillary action.

Natural gas is mostly methane, CH_4. This chemical formula means that one molecule of natural gas contains one atom of carbon and four of hydrogen. When this natural-gas molecule burns, the hydrogen goes to H_2O (water) as before and the carbon to CO_2 (carbon dioxide, two atoms of oxygen and one of carbon).

Q How does hydrocarbon burn?
A There are many hydrocarbon fuels consisting of carbon and hydrogen in various proportions. Whatever the proportions, the hydrogen finally burns to H_2O. The carbon normally burns to CO_2, but sometimes burns incompletely to CO (carbon monoxide).

Even if a cold combustible gas is mixed with exactly the right amount of oxygen for combustion, nothing happens until the temperature of some portion of the mixture is raised to the ignition temperature. This starts the burning, and the heat of the burning gas ignites the next portion of mixture, so that fire spreads rapidly. An example is the cylinder spark in a gas engine which starts the combustion in a small area around the spark gap. Soon the whole mass is aflame.

If a mixture contains too much air, or too little air, it is harder to ignite. It is possible to get the proportions so far from a perfect mixture in either direction that a flame cannot be maintained. Every car driver knows he can be stalled by too rich a mixture as well as by one too lean (a too-lean mixture has a lot of air that can't be used). In either case the dead mass absorbs heat and lowers the temperature produced by combustion. When combustion can't create a temperature high enough for ignition, flame cannot spread in a mixture.

A continuous flame requires a continuous supply of fuel, as in the old-style fishtail gas burner, Fig. 12-2, and in the Bunsen burner, Fig. 12-3. The fishtail gives a yellow flame and the Bunsen a blue flame.

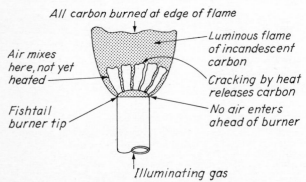

All carbon burned at edge of flame

Air mixes here, not yet heated

Luminous flame of incandescent carbon

Cracking by heat releases carbon

Fishtail burner tip

No air enters ahead of burner

Illuminating gas

FIG. 12-2   Fishtail gas burner shows yellow flame.

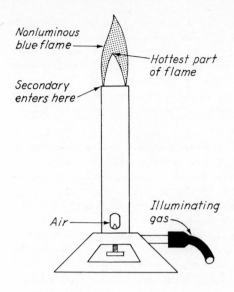

FIG. 12-3 Bunsen burner operates with blue flame.

Q What do yellow and blue flames indicate?
A If pure hydrogen were burned, both flames would be pale blue. Hydrogen cannot produce a yellow flame because it contains no carbon. However, city gas and natural gas do contain substantial quantities of carbon atoms in the molecules of hydrocarbons—compounds of hydrogen and carbon. Here, if the hydrogen can be burned a small fraction of a second ahead of the carbon, the carbon particles are released as a cloud of individual molecules, which can glow brightly for an instant before they are burned. Millions of these incandescent particles constitute a yellow flame.

The fishtail burner soots this gas out, without any contained air, in a thin flat stream through the slotted opening in the tip. The dark area just above the tip is the cold gas in the process of picking up and mixing slightly with the surrounding air. Suddenly, at a certain level, the mixture reaches correct proportion, and starts to burn rapidly. Note that the heat of the flame at any instant is igniting the cooler mixture on the way up.

Q Explain the term *cracking*.
A The temperature splits the hydrocarbon molecule into carbon particles and hydrogen gas. The hydrogen gas burns first with a pale flame, but the flame is colored bright yellow by the millions of incandescent carbon particles. The level at which these particles are completely burned to CO_2 is the top edge of the flame, Fig. 12-2. No further heat is

generated above this point but, of course, the heat previously released in the flame below carries up in the invisible combustion products.

If a cold spoon is held well above the fishtail flame, no soot deposits because all the carbon has been burned. The same spoon stuck into the flame (or into a candle flame) is quickly coated with lamp black (soot). Here the spoon cools the mixture below the ignition point before the carbon is all consumed, so the unburned carbon deposits on the spoon.

Q   How are soot and smoke formed?
A   There is an important point in the previous question for boiler operators. Wherever a yellow flame is allowed to strike relatively cold surfaces (even those as hot as a boiler tube) it will be cooled below the ignition temperature before the carbon is all burned. The unburned carbon deposits as soot on the heating surfaces, or goes up the stack as smoke. In either case, there are waste and pollution.

To prevent smoke and soot, make sure combustion is complete before yellow flames reach the tubes in the boiler. The main ways to do this are to increase furnace height, volume, or temperature, or to insure more perfect mixing so that the combustibles burn up sooner.

Q   Explain how flame behaves in a boiler's furnace.
A   In one area of the furnace, the mixture may be too rich to burn. In other areas there may be almost all air and hardly any combustion gas. It may be so lean a mixture that it burns badly or not at all. Thus, even if the average amount of air is right, the combustion may be bad because some parts of the mixture are too rich and some are too lean.

These explanations may make burning seem simpler than it really is. Actually, the hydrocarbons from coal, oil, and gas may go through a whole series of reactions. Yet the operating engineer must remember that, however it starts, the hydrogen finally ends up in H_2O and the carbon (if it burns fully) in CO_2. But if the carbon burns incompletely, it may produce CO gas, which is capable of further burning, and which is a big waste if allowed to pass up the stack unburned.

Q   What are commercial fuels burned in boilers?
A   All commercial fuels considered in this book consist of one of the following, or of a mixture of two or more of them: (1) gaseous hydrocarbon, (2) solid carbon, or (3) a mixture of carbon monoxide and hydrogen.

Coke and charcoal have little combustible material beyond the solid carbon. Bituminous (soft) coal includes in addition quite a bit of distillate hydrocarbons. Natural gas consists almost entirely of gaseous hydrocarbons. Ordinary carbureted water gas is largely carbon monoxide and hydrogen with a fair amount of hydrocarbons.

Fuel oil vaporizes into gaseous hydrocarbons before it actually burns.

The oil does not burn as oil because the oil is cracked in the flame, thereby producing solid carbon (soot) and hydrogen. Coal is another fuel that decomposes into elementary fuels before actual burning starts. When coal is first exposed to the fire's heat, gaseous hydrocarbons, carbon monoxide and hydrogen are distilled off, leaving solid carbon behind.

Q Explain the actual firing operation in a furnace.

A Broadly speaking, fuel may be burned (1) in suspension or (2) on a fuel bed. The simplest example of suspension burning is gas firing, in which the burner delivers the right proportions of air and gas.

An oil burner first has to convert the oil into a gas. The coal burner must be arranged to distill off the volatile matter soon after the coal enters the furnace. In both cases the furnace heat does the job, but the firing equipment puts the fuel into a condition to make the best use of this heat. For suspension firing of both coal and oil, the condition is the same: the fuel must be broken up into many small particles to expose as much surface as possible.

With oil, this means good atomization. With coal, it means fine grinding. In addition, turbulence is needed to strip away the protective coating of dead gas from the particles.

Q What are the functions of the burner compared with those of the furnace?

A The furnace takes over where the burner leaves off. Within the combustion zone, fuel must be vaporized or distilled, mixed with air, ignited, and the resulting reaction between fuel and oxygen carried to a finish. Mixing fuel and air is chiefly the job of the furnace. Heating and igniting are chiefly the burner's job. We know how hard it is to keep one log burning in a fireplace. With two or more logs the fire keeps going. In the same way, a furnace must maintain a heat supply to prepare and ignite incoming fuel.

Q How are soot and smoke prevented?

A To prevent soot and smoke, the combustion steps (mixing, ignition, turbulence, burning carbon, etc.) must be finished in the short time it takes for the particles to travel from the burner to the furnace outlet. Turbulence is the whirling and eddying of air and combustion gas. For boiler furnaces this is much better than streamlined flow. Turbulence (1) increases the time available for combustion, (2) mixes fuel and air better, and (3) helps the air wash away the dead combustion-product gases and expose fresh surfaces for quick combustion. Without this turbulence, a burning particle soon surrounds itself with a protective layer of unburnable combustion products, mainly CO_2 and nitrogen.

Just remember that good combustion depends upon the three T's: Temperature, Time, and Turbulence. Ash is the part of the coal that can't burn, and hence might not seem to enter this picture. Just the same, it can make trouble. Ash particles, out of control, melt, stick on furnace walls and tubes as slag, wash down refractory walls, and tend to clog the gas passages on their way up to the stack.

Q What are the principal constituents of the fuels used by the power engineer?
A The principal constituents of all fuels used for boiler firing purposes are carbon, hydrogen, oxygen, nitrogen, sulfur, and, for coal, the incombustible elements that form ash.

Q What is the most accurate method of finding the heating value of a fuel?
A The heating value of a fuel can be found most accurately by burning a measured sample with pure oxygen in an apparatus called a *calorimeter,* where the heat of combustion is absorbed in water and the heating value determined by noting the rise in the temperature of the water. The heating value of a solid or liquid fuel is usually given as so many Btu per pound, and the heating value of a gaseous fuel as so many Btu per cubic foot, measured at some standard temperature and pressure.

Q What is a Btu?
A Btu is the abbreviation for *British thermal unit.* It is the amount of heat required to raise the temperature of one pound of water one degree Fahrenheit. At different temperatures the amount of heat required to raise the temperature of one pound of water one degree may be slightly more or less, but the variation is so small that we usually take 1 Btu as the heat required to raise the temperature of one pound of water through one degree at any temperature.

Q Describe a calorimeter for finding the heat values of solid fuels.
A One common form of calorimeter is shown in Fig. 12-4. It consists of a hollow steel cup or *bomb* made in two parts that can be fastened together by a large threaded nut in much the same way as a pipe union. A lead gasket between the faces ensures a tight joint. The walls of the bomb are thick to withstand considerable internal pressure when the fuel sample is burned. It contains a pan for the sample of fuel and means for electrically igniting it. A valve is also provided through which oxygen can be forced into the bomb.

To make a test, a carefully weighed sample of fuel is placed in the pan, the halves of the bomb are screwed tightly together, and the bomb is filled with oxygen under high pressure and then placed in a brass can containing an accurately measured quantity of water at approximately

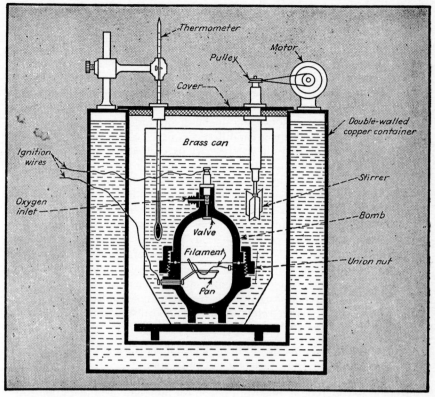

FIG. 12-4 Bomb calorimeter used for finding heat value of fuel.

room temperature. This can is placed inside a double-walled copper vessel containing water, also at room temperature. Sometimes the outer vessel is built like a large vacuum bottle, with the air exhausted from the space between the double walls to provide a better insulating jacket around the inner can. A thermometer in the inner can registers the rise in water temperature, and a small motor-driven paddle wheel stirs the water so that the heat is evenly diffused throughout its entire mass.

At the start of the test, the current is switched on, the glowing wire in the bomb ignites the fuel, and the heat given off by its combustion passes through the walls of the bomb to the water. The rise in temperature of the water is measured by the thermometer. From this reading, together with the weight of the fuel sample burned and the weight of water heated, it is easy to compute the heat value in Btu per pound.

Q What are the principal substances used as fuels by the engineer?
A The principal ones are solids, such as wood and coal; liquids, such as fuel oils derived from petroleum; and gases, such as natural gas, producer gas, blast-furnace gas, and coal gas.

Q Into what classes is coal divided?
A Numerous methods of classifying coal have been used and suggested, based variously on carbon content, ratio between fixed carbon and volatile matter, coking or noncoking qualities, and on other physical characteristics. Nevertheless, the great differences in the composition and physical appearance of coals from the various coal fields, and even from different sections of the same coal seam, make it hard to find any entirely suitable and satisfactory classification system for all coal.

In general there are three main coal classes: anthracite, bituminous, and lignite, but no clear-cut line exists between them, and we have other coals classed as semianthracite, semibituminous, and subbituminous. Anthracite is the oldest coal, geologically speaking. It is a hard coal composed mainly of carbon with little volatile content and practically no moisture. Coming up the scale toward the youngest coal (lignite), carbon content decreases and volatile matter and moisture increase.

Fixed carbon refers to carbon in its free state, not combined with other elements. *Volatile* matter refers to those combustible constituents of coal that vaporize when coal is heated.

Q What are the approximate heat values of the various classes of coal?
A Because of the wide variation in composition it is impossible to give definite figures for each class. Generally speaking, the heat value varies from as low as 7000 to 8000 Btu per lb for lignite to around 15,000 Btu per lb for high-grade bituminous.

Q What is meant by pulverized coal?
A This is coal ground to a fine powder before being fed into the furnace. It is blown into the furnace by a strong blast of air and burns in much the same manner as gas.

Q What are the average composition and heat value of wood?
A Heat value of wood depends largely upon its moisture content. This may run as high as 50 percent in green wood. Even seasoned wood contains 15 to 25 percent moisture—about 8 percent for kiln-dried.

The heat value of perfectly dry wood is about 8500 Btu per lb. The same wood with 25 percent moisture would run about 6400 Btu per lb. Sawdust has much the same heat value as the wood from which it is cut.

Q What are the average composition and heat value of fuel oil?
A Fuel oil may be crude oil direct from the well, but is commonly the

residue left after the lighter oils, such as gasoline and naphtha, have been distilled from the crude. It is composed mainly of hydrocarbons (compounds of hydrogen and carbon) with small quantities of moisture, sulfur, oxygen, and nitrogen. The following is a typical analysis:

	Percent
Carbon	84
Hydrogen	13
Sulfur	1
Oxygen	1
Nitrogen	1

Heat value is approximately 19,000 Btu per lb.

Q What are the composition and heat value of natural gas?
A Natural gas is composed mainly of hydrocarbons, such as methane (CH_4) and ethane (C_2H_6), with small amounts of carbon dioxide (CO_2), oxygen (O_2), nitrogen (N_2), and sometimes hydrogen sulfide (H_2S). The following is a typical analysis:

	Percent
Methane	80.2
Ethane	17.3
Nitrogen	2.3
Oxygen	0.2

The average high heat value is around 1000 Btu per cu ft, measured at 14.7 psi and 60°F.

Q What are the combustible elements in fuels?
A The combustible elements are carbon, hydrogen, and sulfur.

Q What are the noncombustible elements in fuels?
A The noncombustible elements are nitrogen and those elements that make up the compounds forming the moisture and the ash.

Q What element in fuels is harmful in metals?
A Sulfur is harmful because it will combine with condensate to form sulphurous and sulfuric acids (H_2SO_3 and H_2SO_4), which have a corrosive action on iron and steel.

Q What are the products of the complete combustion of carbon, hydrogen, and sulfur?
A If burned with a sufficient supply of oxygen, carbon will burn to carbon dioxide (CO_2), hydrogen will burn to water vapor (H_2O), and sulphur will burn to sulfur dioxide gas (SO_2).

Q What are the products of the incomplete combustion of carbon, hydrogen, and sulfur?
A If the supply of oxygen is insufficient for complete combustion, only part of the hydrogen and sulfur will be burned to H_2O and SO_2; the rest will pass off unchanged. If burned with an insufficient supply of oxygen, carbon will burn to carbon monoxide (CO), instead of CO_2. This carbon monoxide is a combustible gas and is also highly poisonous.

Q What is meant by *excess air,* and why is it necessary?
A The amount of air *theoretically* necessary to burn a fuel completely can be calculated from the ultimate analysis of the fuel, but in practice we must introduce more air into the furnace than the theoretical amount to make sure that all the combustible elements of the fuel receive enough oxygen for complete combustion.

Q What are the usual percentages of excess air admitted to boiler furnaces with (1) coal—hand-fired? (2) coal—stoker-fired? (3) oil, gas, or powdered fuel?
A 1. With hand firing: (*a*) Under good conditions—50 percent; (*b* Under poor conditions—100 percent and over.
   2. With mechanical stokers—20 to 50 percent
   3. With oil, gas, or powdered fuel—10 to 30 percent

Q What are the disadvantages of using large amounts of excess air?
A This excess air does not take part in combustion but absorbs and carries off heat from the furnace to the chimney, thus lowering the efficiency of the steam generating unit.

FUEL ANALYSIS

Q What methods are used to analyze coal?
A There are two methods: the *ultimate analysis* splits up the fuel into all its component elements, solid or gaseous; and the *proximate analysis* determines only the fixed carbon, volatile matter, moisture, and ash percentages. The first must be carried out in a properly equipped laboratory by a skilled chemist, but the second can be made with fairly simple apparatus. Note that *proximate* has no connection with "*approximate.*"

Q Explain how a proximate analysis is made.
A TO DETERMINE MOISTURE: Crush a sample of raw coal until it will pass through a 20-mesh screen (20 meshes per linear inch), weigh out a definite amount, place it in a covered crucible and dry in an oven at about 225°F for one hour. Then cool sample to room temperature and weigh again. The loss in weight represents moisture.

TO DETERMINE VOLATILE MATTER: Weigh out a fresh sample of crushed coal, place in a covered crucible, and heat over a large bunsen burner until all the volatile gases are driven off. Cool the sample and weigh. Loss of weight represents moisture and volatile matter. The remainder is coke (fixed carbon and ash).

TO DETERMINE CARBON AND ASH: Remove the cover from the crucible used in the last test, and heat the crucible over the bunsen burner until all the carbon is burned. Cool and weigh the residue, which is the incombustible ash. The difference in weight from the previous weighing is the fixed carbon.

Q Of what value is a fuel analysis to the engineer?

A It enables the engineer to compare one fuel with another on the basis of moisture, ash, and combustible matter.

SUGGESTED READING

REPRINTS SOLD BY *Power* MAGAZINE
  Coal Selection and Handling, 24 pp.
  Coal Combustion, 24 pp.
  Combustion, Pollution Controls, 24 pp.
  Steam Generation, 48 pp.

BOOKS
  Elonka, Stephen M., and Anthony L. Kohan: "Standard Boiler Operators' Questions and Answers," McGraw-Hill Book Company, New York, 1969.
  Elonka, Stephen M., and Joseph F. Robinson: "Standard Plant Operators' Questions and Answers," vols. I and II, McGraw-Hill Book Company, New York, 1959.

13

FLUE-GAS ANALYSIS, DRAFT, AND DRAFT CONTROL

Flue-gas analysis tells the operator what goes on inside the boiler's furnace so that any excess or deficiency of the needed combustion products can be corrected. With today's automatic controls, combustion-control equipment is more strictly concerned with two basic functions: adjusting the fuel supply to maintain constant steam flow or pressure under varying boiler loads, and correcting and maintaining the ratio of combustion air to the fuel supply.

Whether the boiler is of the natural draft type, or has forced draft equipped with power-operated draft controls, or has fully automatic combustion controls, the operating engineer must see that the exact needed draft is supplied at all times.

FLUE GAS

Q What is flue gas?
A It is the name given to the complete and incomplete products of combustion and excess air passing to the chimney. Commonly considered constituents are water vapor (H_2O), carbon dioxide (CO_2), carbon monoxide (CO), and nitrogen (N_2).

Q What constituents are measured in the ordinary commercial analysis of flue gas?

A Ordinary Orsat apparatus cannot measure moisture. It directly measures the percentages by volume of carbon dioxide, oxygen (O_2), and carbon monoxide in dry flue gas. The remainder is assumed to be nitrogen.

Q What is the flue-gas analysis if carbon is the only combustible element in the fuel and if combustion is complete?

A When carbon burns completely with one volume of oxygen, it produces one volume of carbon dioxide and no carbon monoxide. Note that air contains 21 percent oxygen by volume and 79 percent nitrogen. The nitrogen goes through the furnace unchanged. Any oxygen that burns completely with carbon produces the same volume of carbon dioxide. The remainder of the oxygen is excess and goes through unchanged. Thus, the flue gas here analyzes 79 percent nitrogen, and the oxygen and the carbon dioxide add up to 21 percent.

Thus, if there is no excess air the analysis is

	Percent
Oxygen	0
Carbon dioxide	21
Nitrogen	79
Total	100

If there is 100 percent excess air, unburned oxygen equals that burned to carbon dioxide, and gas analysis is

	Percent
Oxygen	10.5
Carbon dioxide	10.5
Nitrogen	79.0
Total	100.0

If there is 50 percent excess air, the unburned oxygen is half that burned to carbon dioxide, and the gas analysis is

	Percent
Oxygen	7
Carbon dioxide	14
Nitrogen	79
Total	100

Q What is the effect on the flue-gas analysis if the fuel contains no combustible element except carbon, and if part of the carbon is incompletely burned to carbon monoxide?

A The original oxygen now goes in three directions. Part goes through unchanged. Part burns completely to form the same volume of carbon dioxide. The remainder burns incompletely to form twice its volume of carbon monoxide.

Q What is the effect of hydrogen in the fuel?
A Hydrogen burns with part of the oxygen to produce water. This condenses and does not show up in the Orsat analysis, so it is equivalent to a shrinkage.

Q Illustrate the combined effect of all these by an example.
A Suppose that out of the 21 cu ft of oxygen in 100 cu ft of air, 4 cu ft is excess going through unchanged, 1 cu ft burns to make 2 cu ft of carbon monoxide, 13 cu ft burns to 13 cu ft of carbon dioxide, and 3 cu ft burns with hydrogen to produce water. The 79 cu ft of nitrogen goes through unchanged.

The first column below gives the products of combustion in cubic feet, and the second column in percent:

	Cu ft	Percent
Oxygen	4.0	4.1
Carbon monoxide	2.0	2.0
Carbon dioxide	13.0	13.3
Water (disappears)	0.0	
Nitrogen	79.0	80.6
Total	98.0	100.0

In this problem, carbon monoxide is purposely shown higher than would normally occur in boiler operation. Generally the figure is less than 1 percent.

FLUE GAS ANALYSIS

Q Sketch and describe an apparatus for analyzing flue gases.
A The Orsat apparatus, Fig. 13-1, is a common type of flue-gas analyzer. Its principal parts are a water-leveling bottle, a water-jacketed measuring burette, three pipettes, three containers, various connecting tubes and valves, and a frame to hold all these parts rigidly in position.

The leveling bottle and measuring burette contain pure water. The first pipette (a) is packed with steel wool kept wet with caustic-potash solution from the container below. The steel wool in the second pipette (b) is wet with pyrogallic solution from the lower container. The third pipette (c) contains copper strips wet with cuprous (copper) chloride solution.

Pipettes (a), (b), and (c) are for the absorption of CO_2, O_2, and CO, in the order named. Other chemicals than those indicated are sometimes used for these absorptions.

The Orsat is connected to the sampling tube in the stack or breeching by a length of rubber tubing in series with an aspirator rubber bulb, Fig. 13-2, to pump the gas and (generally) a gas filter (glass wool in a glass tube) to remove dust and tar. The gas line connects to the back side (not shown, Fig. 13-1) of the three-way cock. This cock has three positions: (1) to connect sampling tube to measuring burette; (2) to connect burette to atmosphere; (3) to connect sampling tube to atmosphere. (This last position is *closed* as far as the Orsat is concerned.)

Q Explain how a flue-gas analysis is made, using the indicated type of Orsat.

A The following covers the main steps only. For actual operation always study detailed instructions supplied with the particular instrument. These show how to avoid contamination of the sample with air or of the chemical with water, or vice versa—also how to insure easy operation and precise readings.

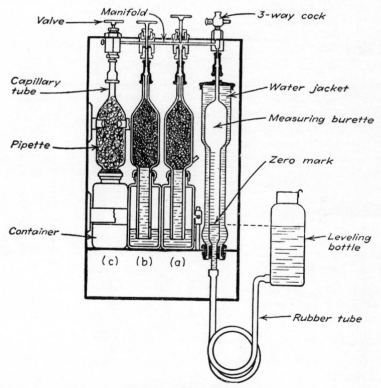

FIG. 13-1   Orsat flue-gas analyzer is portable unit.

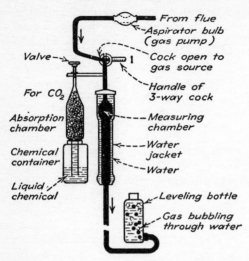

FIG. 13-2 Pumping sample of flue gas into measuring burette.

STEP 1 (Fig. 13-2): Use hand aspirator bulb to pump gas through measuring burette. It bubbles out of measuring bottle as shown. Continue until sure that all air has been displaced.

STEP 2 (Fig. 13-3): With three-way cock set to connect burette to atmosphere, raise leveling bottle slowly until water is at zero mark near bottom of burette. This is also level with water in bottle, so sample is measured at atmospheric pressure. (All gas measurements must be made at atmo-

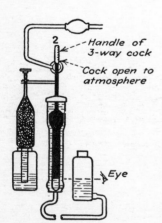

FIG. 13-3 Bringing gas sample to zero mark at atmospheric pressure.

spheric pressure, that is, with water level the same in burette as in leveling bottle.)

STEP 3 (Fig 13-4): To absorb CO_2, close three-way cock and lift bottle high. Open valve to CO_2 pipette, so gas displaces liquid and contacts caustic solution on steel wool, which absorbs CO_2. Pinch tube to check flow when water rises to mark in capillary tube above burette. Allow a few seconds for CO_2 absorption.

STEP 4 (Fig. 13-5): Next, keeping tube pinched, lower the bottle. Then release rubber tube cautiously until liquid rises exactly to the mark in capillary tube on top of pipette. After closing pipette valve, raise or lower bottle until water level in bottle is exactly the same as in burette. Read graduation at water level in burette. This is CO_2 percentage. Repeat steps 3 and 4 to make sure all CO_2 is absorbed.

STEP 5: Repeat steps 3 and 4 for pipette (*b*) to absorb the O_2. The *increase* in burette reading is percentage of O_2.

STEP 6: Again repeat steps 3 and 4 with pipette (*c*) to absorb the CO. The *further* increase in burette reading is the percentage of CO.

Q What are some of the peculiarities of the chemicals used in the Orsat?
A The caustic to absorb CO_2 acts surely and rapidly. This solution need not be protected from the air and lasts a long time. The CO_2 should always be absorbed first.

The solutions for absorbing O_2 and CO act more slowly, and must be protected from contact with the air. If the O_2 is not completely absorbed

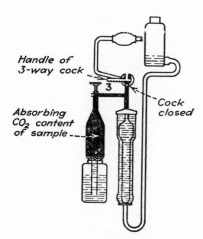

FIG. 13-4 Transfering gas sample to first absorption pipette.

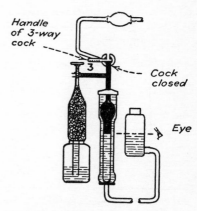

FIG. 13-5 Returning gas sample to burette to measure percentage of CO_2.

in pipette (*b*), it will be in pipette (*c*) and will therefore be erroneously reported as part of the CO. Be sure O_2 is *completely* absorbed in (*b*).

If a second pass through any pipette produces no increase in absorption, it may generally be assumed that the solution is in good shape.

Q Give an example of readings during an Orsat analysis.

A Reading after CO_2 absorption, 12.8 percent; after O_2 absorption, 18.7 percent; and after CO absorption, 19.8 percent. Then analysis is

	Percent
CO_2	12.8
$O_2(18.7 - 12.8)$	5.9
$CO(19.8 - 18.7)$	1.17

Q What gas analysis results are reasonable?

A The following comments hold only for solid or liquid fuels (not for gas fuels): The sum of the CO_2, O_2, and half the CO should never be above 21. With coal this sum should never be less than 18.5. With fuel oil, the sum should never be less than 15.5.

Q Assuming there is no CO in the gas, give a table showing the percentage of CO_2 for various percentages of excess air with typical anthracite coal, bituminous coal, fuel oil, and natural gas. Also show for each CO_2 percentage about what the O_2 should be, as a check on the analysis.

A See Table 13-1.

Q Which is the more reliable indicator of excess air, CO_2 or oxygen?

A Table 13-1 shows that oxygen percentage is the more reliable indicator, because it reads about the same for any coal, and almost the same

TABLE 13-1 Corresponding Percentages of Excess Air, Carbon Dioxide, and Oxygen

		Percent excess air					
Kind of fuel	Ingredients	0	20	40	60	80	100
Anthracite	CO_2	19.5	16.0	13.8	12.0	10.7	9.6
	O_2	0.0	3.5	6.1	8.0	9.4	10.6
Bituminous	CO_2	18.6	15.5	13.2	11.5	10.1	9.2
	O_2	0.0	3.5	6.0	8.0	9.5	10.6
Fuel oil	CO_2	15.5	12.6	10.6	9.3	8.2	7.4
	O_2	0.0	3.7	6.4	8.1	9.6	10.8
Natural gas	CO_2	12.2	10.0	8.5	7.5	6.5	5.7
	O_2	0.0	4.0	6.5	8.5	9.9	10.9

NOTE: Where gas contains CO, substitute for CO_2 the sum of the CO_2 and half the CO. Thus, if CO_2 is 12% and CO is 2%, equivalent CO_2 is 13%.

for oil as for coal. For any coal or fuel oil the excess air can be closely estimated from the O_2 by the following rule.

Subtract the O_2 percentage from 21. Divide the O_2 percentage by this difference and multiply by 100 to get the percentage of excess air.

Example: What is the percentage of excess air for coal or oil if the O_2 in the flue gas is 8.3 percent? Solution: $21 - 8.3 = 12.7$. Then $8.3 \div 12.7 = 0.65$ and $0.65 \times 100 = 65$ percent excess air. *Ans.*

Q Give some typical flue-gas analyses for bituminous-coal firing under various conditions.

A Analyses are as follows:

	% CO_2	% O_2	% CO	% N_2 (by difference)
Poor	9	10.5	0.5	80.0
Average	12	7.5	0.1	80.4
Good	15	4.0	0.2	80.8

Note (from Table 13-1) that the last condition above corresponds to less than 20 percent excess air. Any further reduction in excess air would undoubtedly raise the CO excessively.

Q What percentage of CO_2 do you expect when burning natural gas under good conditions?

A Because of the high H_2 content of natural gas, 20 percent excess air, which is about the practical minimum, corresponds to about 10 percent CO_2.

Q What percentage CO_2 would you expect when burning fuel oil under good conditions?

A The C content is relatively higher in fuel oil than in natural gas, and so the attainable percentage of CO_2 will be higher (but less than with coal). With 20 percent excess air, CO_2 with oil fuel should be about 12.5 percent.

Q What is a CO_2 recorder? Explain the type shown in Fig. 13-6.

A CO_2 recorder is an apparatus for automatically determining the percentage of CO_2 in flue gas and recording this analysis on a chart. In Fig. 13-6, a small stream of water passing through the aspirator draws a sample of flue gas into the measuring burette (4), and at the same time fills the standpipe containing the burette. When the water in the standpipe reaches the level marked V, a quantity of gas is trapped in the measuring burette. As the water continues to rise, the surplus gas passes off through the atmospheric tube until it is cut off at *W*, leaving an accurately measured sample of flue gas in the burette, under atmospheric pressure.

The water continues to rise in the standpipe, forcing the sample of gas over into the absorption chamber (*9*), which is filled with steel wool saturated with a caustic solution (*cardisorber*). As the gas comes in contact with the wet surfaces of the steel wool, the CO_2 is absorbed. At the same time some of the liquid in the absorption chamber is forced back into the caustic tank, raising the level of the liquid in the compression cylinder.

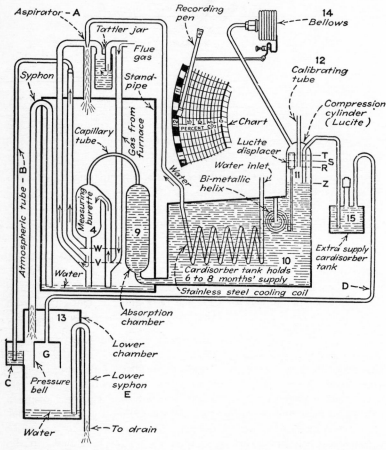

FIG. 13-6 CO_2 recorder for automatically measuring percentage of CO_2 in flue gas.

When the liquid level in the compression cylinder reaches R, the calibrating tube is sealed off, and any further rise of the liquid forces air into the bellows (*14*). The expansion of the bellows moves the recording pen over the chart a distance proportional to the amount of gas left in the sample after the CO_2 has been absorbed.

While this is going on, the water keeps rising in the standpipe until it reaches a level high enough to start the siphon flowing. The siphon then empties the standpipe. The pressure on the caustic in the absorption chamber being now released, the solution rises to its original level in the absorption chamber, forcing the remainder of the gas sample back into the measuring burette. A stream of fresh flue gas sweeps through the burette, clearing out the old gas. Then the new sample is trapped, and the process is repeated. The flue gas is cleaned by bubbling it through water before it enters the recorder.

Q Does a high percentage of CO_2 always indicate good combustion?
A High percentage usually indicates that the excess air is being kept to the minimum, but it does not necessarily indicate the best attainable combustion, as there may also be a considerable percentage of CO. A mere 1 percent of CO may mean a loss of nearly 5 percent of the total heat value of the fuel burned. CO_2 should be high, with little or no CO.

Even when a CO_2 recorder is used, gas should be checked periodically with an Orsat to see whether CO is present.

Q Give a formula for estimating the approximate heating value of a fuel from the ultimate analysis.
A Dulong's formula, shown below, is probably the most widely used for this purpose. It is based on the known heat values of 1 lb each of pure carbon, pure hydrogen, and pure sulfur. Each of these values is multiplied in turn by the percentage of that particular element in the fuel, and the results are added together.

In the formula, the heat value of 1 lb of hydrogen is taken as 62,000 Btu, the heat value of 1 lb of carbon as 14,600 Btu, and that of 1 lb of sulfur as 4000 Btu. It is assumed that all the oxygen shown in the analysis is already combined with one-eighth its weight of hydrogen in the form of moisture, and the hydrogen so combined is therefore incombustible. The symbols C, H, O, and S represent the percentages of carbon, hydrogen, oxygen, and sulfur. The formula is as follows:

$$\text{Heat value per pound} = \frac{14{,}600C + 62{,}000\left(H - \dfrac{O}{8}\right) + 4{,}000S}{100}\ \text{Btu}$$

Q Give a formula for estimating the theoretical amount of air required for the complete combustion of 1 lb of fuel.

A The following formula is based on the known amount of air required for the complete combustion of 1 lb each of pure carbon, pure hydrogen, and pure sulfur. In the formula the figures represent this air in pounds, and the symbols are the percentages of carbon, hydrogen, oxygen, and sulfur.

$$\text{Air required, pounds} = \frac{11.61C + 34.8\left(H - \dfrac{O}{8}\right) + 4.35S}{100}$$

Q Calculate the heat value of a coal having the following ultimate analysis: C, 82.4 percent; H, 4.1 percent; O, 2.3 percent; S, 0.5 percent; ash, 10.7 percent.

A From Dulong's formula:

$$\text{Heat value} = \frac{14,600 \times 82.4 + 62,000\left(4.1 - \dfrac{2.3}{8}\right) + 4000 \times 0.5}{100}$$

$$= 146 \times 82.4 + 620 \times 3.81 + 40 \times 0.5$$
$$= 12,030 + 2362 + 20$$
$$= 14,412 \text{ Btu} \quad Ans.$$

Note that the heating value of the sulfur is negligible.

Q Calculate the theoretical amount of air required for the complete combustion of the fuel in the previous question.

A Air required $= \dfrac{11.61 \times 82.4 + 34.8\left(4.1 - \dfrac{2.3}{8}\right) + 4.35 \times 0.5}{100}$

$$= \frac{11.61 \times 82.4 + 34.8 \times 3.81 + 4.35 \times 0.5}{100}$$

$$= \frac{957 + 133 + 2}{100}$$

$$= 10.9 \text{ lb of air per lb coal} \quad Ans.$$

Q For coal firing, describe the various steps in the combustion process from the time fresh coal is thrown on the fire.

A 1. If the coal contains moisture, it is driven off.

2. The dry coal absorbs heat, and the volatile constituents begin to distill off.

3. The volatiles mix with air and burn, their products of combustion passing up the chimney.

4. The fixed carbon left on the grate is now burned until only the ash remains.

Q What furnace conditions are essential to burn coal efficiently and economically?
A 1. Enough air must be admitted to supply the oxygen necessary for the complete combustion of the combustible elements of the fuel.

2. Admission of air must be regulated in such a way that it mixes thoroughly with the combustible gases distilled from the coal and comes in contact with all the carbon left on the grate after the gases are driven off and burned.

3. Temperature of the furnace must be kept at not less than 1800°F.

NATURAL AND MECHANICAL DRAFT

Q What is *draft*?
A *Draft* is the difference of pressure producing air flow through a boiler furnace, flue, and chimney.

Q Why is draft necessary?
A It supplies air in sufficient *quantity* to ensure complete combustion and under sufficient *pressure* to overcome the resistance offered by the boiler shell, tubes, furnace walls, baffles, dampers, breeching, and chimney lining.

Q In what units is draft measured?
A It is measured in *inches of water column* as indicated by (1) difference between the water-column heights in the two legs of a glass U tube; (2) height of the column in an inclined-glass draft gage; (3) pointer reading in an indicating or recording draft gage.

Q How is draft measurement in inches of water converted into pressure in pounds or ounces per square inch?
A Weight of a cubic foot of water varies with temperature, being 62.43 lb at the point of maximum density, 39.2°F. The drop in weight is very slight over the ordinary range of boiler-room temperatures, so for all practical purposes take it as 62.4 lb per cu ft. Since 1 cu ft = 1,728 cu in., dividing 62.4 by 1,728 gives 0.036 psi, the pressure in pounds per square inch indicated by a column of water 1 in. high. This number multiplied by 16 gives 0.576, the pressure in ounces per square inch.

Q Sketch a U-tube draft gage, and explain its operation.
A Figure 13-7 illustrates the principle of the U-tube draft gage (also called a manometer). One end of the glass U tube is open to the atmosphere, the other connected to the boiler furnace, flue, chimney, or other enclosed space where it is desired to measure draft pressure. The U tube is partly filled with water, and when there is no difference in the pressure

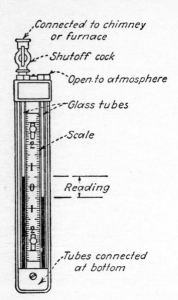

FIG. 13-7 U-tube draft gage.

acting upon the water surface in the two legs, water level is the same in each leg and the reading on the scale zero.

If the pressure within the boiler setting or chimney is *greater* than the atmospheric pressure, it forces the water down in the leg connected to the setting, and up a corresponding amount in the leg open to the atmosphere. If the pressure within the boiler setting is *less* than the atmospheric pressure, the latter forces the water down in the leg open to the atmosphere and up in the leg connected to the setting. The difference in inches between the levels of the two water columns is the draft measurement. In Fig. 13-7 the scale is divided into inches and tenths of an inch, and the water levels show a reading of 1 in. of draft.

Figure 13-8 shows a more elaborate U-tube gage. Oil is used instead of water, and the scale is calibrated to read directly in hundredths of inches of water, with the aid of a vernier. Connected as shown, the scale reads pressure above atmospheric. Suction can be read by changing the top connections.

When taking a reading on this, or any other instrument where the water level in a tube must be read, note that the water surface is not flat, but concave. The curvature is caused by surface tension, and the shape assumed by the water is called a *meniscus*. The same phenomenon is observed with lighter liquids, such as oils, but when mercury is used the

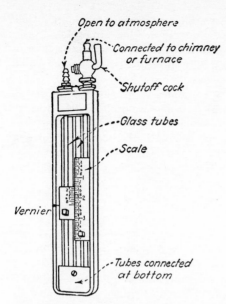

FIG. 13-8   U-tube draft gage with vernier for more accuracy.

meniscus is reversed, or convex. The bottom of the meniscus is usually read on the scale when water or oil is used and the top when mercury is used.

Q Sketch an inclined-tube draft gage and explain its operation.

A This gage is the same in principle as the U tube, but one leg is inclined (Fig. 13-9), and a liquid is used that will not evaporate so readily as water. As the liquid moves through a considerable distance in the inclined tube for a very small change in vertical height, a much larger scale can be used

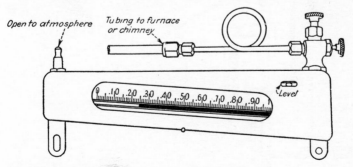

FIG. 13-9   Inclined-tube draft gage must be level.

than that in the vertical-tube gage, so that readings are much finer. The gage scale in Fig. 13-9 is registering pressure below atmospheric pressure, the reading being 0.25 in. of draft. Here, the furnace pressure is below atmospheric, and the atmospheric pressure is forcing the liquid down in the inclined tube.

With chimney draft alone, or with chimney plus induced-draft fan alone, the furnace pressure is always less than atmospheric. If a forced-draft fan is added, pressure in the furnace may be greater or less than atmospheric, or equal to it—that is, balanced. If furnace pressure is higher than atmospheric, the connection shown causes the liquid to be forced upward in the inclined tube. Therefore either the scale or the connection must be reversed.

Q Describe a diaphragm-operated draft gage.
A Figure 13-10 shows the operating principle of this gage. A thin diaphragm of very flexible material is enclosed in a metal case and connected through suitable links or gearing to a pointer that moves over a scale graduated to read in inches of draft. Referring to the sketch, the compartment to the right of the diaphragm is airtight and that to the left is open to the atmosphere. The airtight compartment is connected by tubing to the point where it is desired to measure draft pressure. The movement of the diaphragm under this pressure is transmitted to the pointer by the connecting links so that it moves over the scale and indicates the pressure.

When the pressure to be measured is below atmospheric, the same instrument can be used, but the tubing is connected to the left-hand com-

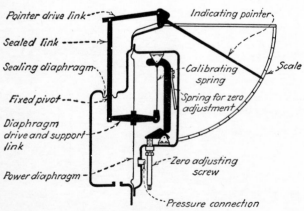

FIG. 13-10 Diaphragm-operated draft gage.

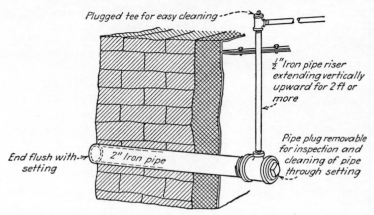

FIG. 13-11 Draft-gage connection to furnace.

partment, and the right-hand compartment is left open to the atmosphere.

Recording draft gages act on the same principle as indicating gages, but the end of the pointer carries a pen which is in contact with a moving chart.

Metal expanding bellows, bourdon tubes made in single-tube and spiral- and helical-tube forms, sealed oil columns, and mercury columns operating floats, are also used as pressure elements in indicating and recording draft gages.

Q Where is draft usually measured, and how are draft gages connected to the furnace or gas passages?
A Gages may be connected and draft measured at any point between the ashpit and the chimney bottom. Where gages are permanent, use at least two, one showing draft over the fire and one the draft at the last pass of the flue gases. Figure 13-11 shows one method of connecting a draft gage to a brick setting. A pipe is inserted in the wall and cemented to prevent leakage. A plugged tee is placed on the outer end of this pipe to provide a connection for the draft-gage tubing and a means of cleaning the large pipe if it should become plugged with soot or ashes.

Q What is *natural* draft?
A It is draft produced by a chimney alone. It is caused by the difference in weight between the column of hot gas inside the chimney and a column of cool outside air of the same height and cross section. Being much lighter than outside air, chimney gas tends to rise, and the heavier outside air flows in through the ashpit to take its place.

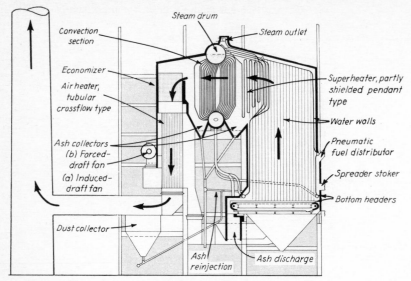

FIG. 13-12 Induced and forced draft produced by fans moving air through furnace.

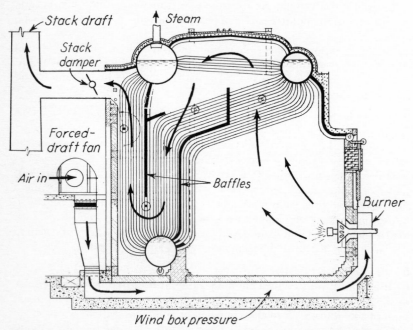

FIG. 13-13 Forced draft pushes air into wind box, where it is preheated before entering furnace.

Q How is natural draft controlled?
A Usually by hand-operated dampers in the chimney or breeching connecting the boiler to the chimney.

Q What is *mechanical* draft?
A It is draft artificially produced by mechanical devices, such as fans and, in some units, steam jets.

Three basic methods of applying fans to boilers are:

1. Balanced draft where a forced-draft (F-D) fan (blower), Fig. 13-12, pushes air into the furnace and an induced-draft (I-D) fan (draws) or a high stack (chimney) provides draft to remove the gases from the boiler. Here the furnace is maintained at from 0.05 to 0.10 in. of water-gage below atmospheric pressure.

2. An induced draft fan or the chimney provides enough draft for flow into the furnace, causing the products of combustion to discharge to atmosphere. Here the furnace is kept at a slight pressure below the atmosphere so that combustion air flows through the unit.

3. The pressurized furnace, Fig. 13-13, uses a blower to deliver the air to the furnace, causing combustion products to flow through the unit and up the stack.

Fans are also used to supply over-fire air, Fig. 13-14, to the furnace in some designs. Pulverized-fuel-burning boilers have blowers which deliver the coal dust to the furnace.

Q What are the main characteristics of *forced* draft?
A *Forced* draft often (but not necessarily) maintains a pressure above atmospheric in the boiler setting; therefore flue gas may be forced out into the boiler room through cracks or leaks in the setting. Such leakage should not be important with a boiler setting in good condition. In a few rare cases where boilers are hand-fired, flame may shoot out of firing doors if the draft is not cut out before the doors are opened.

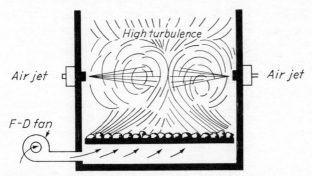

FIG. 13-14 Air jets provide turbulence over fuel bed.

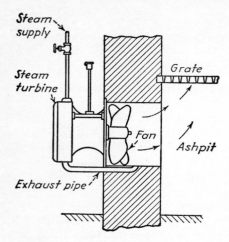

Steam supply

Steam turbine

Grate

Fan

Ashpit

Exhaust pipe

FIG. 13-15 Forced draft using turbine blower beneath fuel bed.

Q Sketch an F-D fan driven by a turboblower.

A Figure 13-15 is a forced-draft propeller fan built into the side wall of a boiler setting. A small direct-connected steam turbine fed by steam from the boiler assures draft in case of electric power failure.

Forced draft is commonly used with underfeed stokers, since considerable pressure is needed to force the air required for combustion up through the deep fuel bed. In general, all types of large boilers operated at high ratings are normally equipped with forced draft.

Forced-draft fans may draw their air supply from the boiler room or from air preheaters, the latter being widely used with large boiler units.

Q What are the main characteristics of *induced* draft?

A *Induced* draft creates a partial vacuum in the boiler furnace so there is no likelihood of furnace gas leaking out through cracks in the setting. If any cracks are present, air is drawn in and it may reduce furnace efficiency considerably. Because of this, induced draft may not be as satisfactory as forced draft where solid fuel is fired and the fuel bed is thick, but it may fit the requirements of small or medium-sized plants where furnaces are hand-fired, equipped with overfeed stokers of certain types, or where boiler fuel is oil or gas.

Induced-draft fans are of similar construction to forced-draft fans, but are usually of the single inlet type, and larger than forced-draft fans for the same boiler output, because the hot flue gas is much greater in volume than the air originally fed to the furnace. Induced-draft fans are also likely to deteriorate more rapidly than forced-draft fans as they are exposed to higher temperatures, cinders, and possible corrosive action.

Q What is *balanced* draft?

A This term is usually applied to a combination of forced and induced draft, automatically regulated to keep the boiler furnace at approximately atmospheric pressure.

AUTOMATIC DRAFT CONTROL

Q Discuss *automatic draft control.*
A It is the automatic regulation of fans and dampers for increasing or decreasing air flow to maintain constant steam pressure as the load changes, and also to maintain good combustion conditions. Devices to control draft-fan speed and damper position frequently employ diaphragms. Changes of air or steam pressure act upon the sensitive diaphragms to open or close the electrical switches, fluid valves, or steam valves that control fan speed or damper position.

Q Explain a simple form of draft control.
A Figure 13-16 shows a simple form of draft control in which the furnace pressure acts upon an enclosed metal diaphragm, and the consequent movement of this diaphragm in turn moves a balanced double

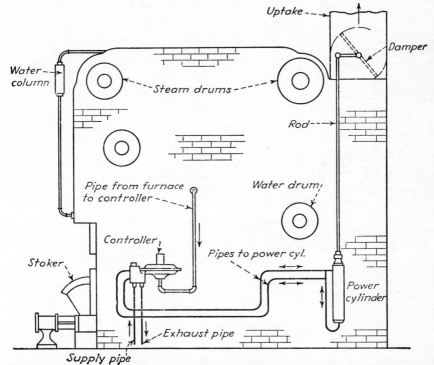

FIG. 13-16 Power operated damper control is fully automatic.

valve, controlling the flow of fluid under pressure to a cylinder containing a piston, one end of which is attached by rods or chains to the uptake damper.

The action of the control valve is to admit fluid under pressure above or below the piston in the power cylinder and at the same time release the fluid pressure on the opposite side of the piston so that the piston can move up or down and in doing so, open or close the damper and hold it in that position until a further change in draft pressure takes place.

If the hydraulic pressure is obtained from some convenient water line, the escaping water from the power cylinder can drain to waste, but if oil is being used, it drains back to a storage tank from whence it is pumped back by a pump arranged to keep a constant pressure on the supply line to the control valve.

Q Explain operation of balanced-draft control.
A In Fig. 13-17 the fan speed is controlled to burn coal fast enough to

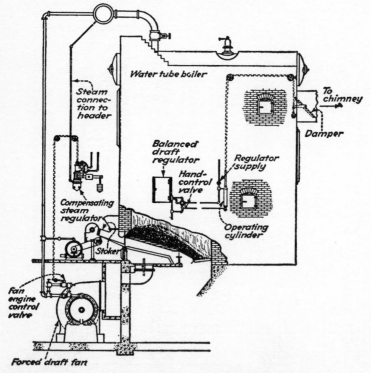

FIG. 13-17 Balanced-draft installation for stoker-fired boiler.

hold the steam pressure constant at all loads. The damper's position is adjusted to "balance" the draft—that is, to maintain practically atmospheric pressure in the furnace.

The balanced-draft regulator is a large diaphragm exposed on one side to furnace pressure conveyed through a draft tube built into the wall of the setting. Movement of this diaphragm operates a pilot valve that admits water or air under pressure to the damper-control cylinder, causing the piston to adjust the damper. When the damper takes up its new position, the furnace pressure comes back to normal, and the diaphragm returns to its original position, no further movement taking place until another change occurs in the furnace pressure. The action of the diaphragm under furnace pressure is opposed by springs that can be adjusted to secure any desired pressure within the range of the regulator.

Fan speed is controlled by a steam-pressure regulator operated by rise or fall of steam pressure in the steam main. If steam pressure falls a little, the regulator opens the fan engine's throttle valve to speed up the fan sufficiently to carry the higher load.

Q Explain the use of both forced and induced draft for one boiler shown in Fig. 13-18.

A In Fig. 13-18 air is forced into the ashpit by a forced-draft fan, and the waste furnace gases are drawn through an economizer and discharged to the chimney by an induced-draft fan. The boiler feedwater passes through rows of tubes in the economizer and absorbs some of the heat from the waste flue gases before they go to the chimney.

Q What advantages has mechanical draft over natural draft?

A Mechanical draft is independent of wind or temperature changes. After the limit of combustion is reached with natural draft, more fuel may be burned and the boiler capacity increased by installing mechanical draft; also, poorer grades of fuel may be burned than with natural draft. Where a very thick fire is carried, as for underfeed stokers, mechanical draft must be used to force air through the thick fuel bed. Also, when economizers are placed between the boiler and chimney to utilize heat from the waste flue gas, mechanical draft is essential to overcome the resistance offered by the banks of tubes in the economizer.

CHIMNEYS

Q Why is a chimney necessary?

A It creates natural draft where no draft fans or blowers are used and discharges the products of combustion at such a height that they will not be a nuisance to the surrounding community. Where mechanical draft is used the second reason is the main one.

Q What are the common types of chimneys, and the advantages and disadvantages of each?

A Chimneys are built of steel, brick, or concrete. The steel chimney is usually the cheapest and most easily erected, but it requires more care and attention if it must last a comparable time. Unless the outside is kept well painted, a steel stack deteriorates rapidly from the action of the weather, and the flue gas corrodes the inner surface if it has no protective lining. Also, the excessive heat loss through an unlined steel stack reduces the draft produced; hence practically all permanent steel stacks are lined with firebrick or some other fire-resisting material. A common practice in building this lining is to use firebrick for the base section, where the heat is most intense, and common brick for the rest of the distance to the top.

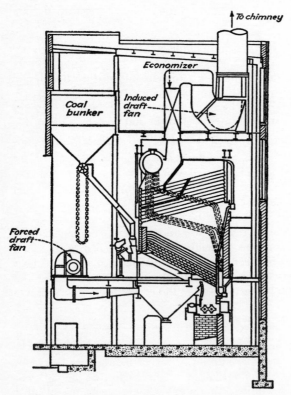

FIG. 13-18   Boiler using forced and induced draft.

Steel chimneys may be self-supporting, in which case the base flares out to about twice the upper chimney diameter and is bolted to a substantial concrete foundation. If the chimney is not self-supporting it is kept upright, and braced against wind pressure, by steel guy wires or cables with one end anchored in the ground and the other fastened to a ring on the chimney, about two-thirds of the height from the bottom. Small steel stacks are usually of uniform diameter throughout, but large stacks, especially the self-supporting type, may be tapered gradually from bottom to top.

The brick chimney is more expensive to build than the steel, but it lasts longer and stands weathering much better. It is usually built with a uniform inside diameter from bottom to top but a decreasing outside diameter. In the best construction, an inner lining is used with an annular space between the lining and the outer wall, so that the lining expands or contracts without affecting the outer wall in any way. A hard close-grained brick is used for the outer-wall construction. The inner lining may be all firebrick, or firebrick part way up from the base and common brick the rest of the way. Rectangular bricks can be used for chimney building, but specially shaped radial bricks make a stronger and neater job.

Chimneys are also built of reinforced concrete. This type can be built rapidly, yet it is very strong because of the steel reinforcing. The chimney is practically airtight and less likely to have air leakage than a brick chimney with its multitude of joints. Walls may be thinner than brick, thus giving a lighter construction without sacrificing strength. A concrete chimney should also have an inner refractory lining at least part of the way up from the base.

Q Numerate the principal points to observe when designing a chimney.
A They are:
1. Height must give the desired draft.
2. Cross-sectional area must be sufficient for boiler load served.
3. Foundation must be solid and substantial.
4. Base must support the weight.
5. Chimney must resist maximum wind pressure to which it is likely to be exposed.
6. It must have high resistance to the weathering action of wind, rain, heat, and cold.
7. A good inner lining should protect the inside of the chimney and prevent excessive heat loss through radiation.

Q Why are chimneys lined, and what materials are used?
A All types should be lined to prevent damage from the effects of

unequal expansion due to the difference between flue-gas temperature inside the chimney and that of the outside air. The lining also protects against the corrosive action of the products of combustion passing up the chimney and can be renewed without rebuilding the entire chimney. A space should be left between the lining and the main wall, and the lining should not be bonded to the outer wall. The chimney may be lined completely or only part way up. Concrete, common brick, firebrick, and other refractory materials are used for chimney lining.

Q Give a formula for calculating the draft that can be produced by a chimney of a given height.

A The following formula is commonly used. It is based on the fact that air and flue gas expand in volume with increase of temperature, so that the higher the temperature the less they weigh per cubic foot. The formula for the resulting natural draft is

$$D = 0.52 \times H \times P \left(\frac{1}{T_o} - \frac{1}{T_c} \right)$$

where D = draft pressure, in. of water
H = height of chimney, ft
P = atmospheric pressure, psi absolute
T_o = absolute temperature of outside air, °F
T_c = absolute temperature of chimney gas, °F
NOTE: Absolute temperature is Fahrenheit temperature $+460°$.

Q What draft is produced by a chimney 100 ft high, temperature of chimney gas 500°F, outside temperature 60°F, atmospheric pressure 14.7 psi?

A $T_o = 60 + 460 = 520$
$T_c = 500 + 460 = 960$

$$D = 0.52 \times H \times P \left(\frac{1}{T_o} - \frac{1}{T_c} \right)$$

$$= 0.52 \times 100 \times 14.7 \left(\frac{1}{520} - \frac{1}{960} \right)$$

$$= 764(0.00192 - 0.00104) = 764 \times 0.00088$$

$$= 0.67 \text{ in.} \quad Ans.$$

Note that this is not the *net* draft the chimney would produce with a boiler carrying a load. Instead it is the *static* draft; that is, the draft the hot gas in the stack would produce if there were no flow of gas up the stack. In actual operation the gas encounters resistance flowing up the stack, and this uses up part of the total draft as computed by the formula just given. This draft loss in the stack increases with the rate of fuel burning and is much greater for a small-diameter than for a large-diameter stack. To keep it within reasonable limits the cross-sectional area of the chimney opening must be increased almost in proportion to the amount of coal burned by the boilers it serves.

Q Is a higher or a lower chimney needed when burning oil or gas fuel than when burning coal?
A Use of oil or gas fuel eliminates the resistance of grates and fuel bed, and hence permits a lower chimney unless (and this is important) the stack must allow for changing over to coal. Excessive draft from a stack higher than necessary can be reduced by dampers.

Q What is meant by *total* draft pressure and *available* draft pressure?
A The draft formula in two previous questions calculates the *total* static-draft pressure. The *available* draft is the draft at any point from the ashpit to the chimney top that is available after the *pressure drop* to that point has been deducted from the total static-draft pressure.

The total available draft is equal to the total theoretical static draft, less the pressure used in producing velocity and overcoming friction within the chimney and connecting passages.

Q Give a formula for finding the height of a chimney to produce a given total or static draft.
A With the previous formula transposed,

$$H = \frac{D}{0.52P \left(\dfrac{1}{T_o} - \dfrac{1}{T_c} \right)}$$

Q What chimney height is required to give a total static draft of 0.8 in. of water column with a boiler-room temperature of 90°F, average temperature of chimney gas 500°F, and atmospheric pressure 13 psi?

A $T_o = 90 + 460 = 550$
$T_c = 500 + 460 = 960$

$$H = \frac{0.8}{0.52 \times 13 \left(\dfrac{1}{550} - \dfrac{1}{960} \right)}$$

$$= \frac{0.8}{0.52 \times 13(0.00182 - 0.00104)}$$

$$= \frac{0.8}{0.52 \times 13 \times 0.00078} = \frac{0.8}{0.00527} = \frac{800}{5.27}$$

$$= 152 \text{ ft} \quad Ans.$$

Q How do you find the chimney area for a given height and coal consumption?
A There are many different formulas for finding chimney area, but all are of a rule-of-thumb nature and give undependable results. The type of chimney construction, method of firing, boiler and furnace design, etc., are among the variable factors that make it impossible to evolve a single formula to suit all conditions.

The only safe procedure is to turn over the actual determination of chimney dimensions, particularly the internal diameter, to an engineer who is able to calculate them directly from a knowledge of the engineering principles involved, and who has full knowledge of the boiler and furnace design, baffling, firing method, load conditions, etc.

MAINTAINING A CLEAN STACK

With strict air pollution regulations in force today, the fireman's job is to maintain a "clean stack." A "dirty stack" not only pollutes the air, but wastes fuel by sending unburned carbon and other particulates up into the atmosphere and may result in a heavy fine to the plant owner.

Fly ash and unburned carbon particles are the principal emissions from coal-fired boilers. Some of the more objectionable particulates are nitrogen oxide (NO_x) and sulfur dioxide (SO_2). For removing some of these, cyclone collectors, filters, electrostatic precipitators and venturi scrubbers are used. We cannot give details here, but the boiler room operator should know how they work and also how to keep them in repair. To burn fuel efficiently, the person in charge of a boiler plant must first of all know what causes a dirty stack.

Q What causes heavy black smoke when fuel oil is burned?
A Insufficient air or excess fuel causes smoke. As air needed for combustion is increased, the smoke lightens until it turns into a light-brown haze; this corresponds to the best operating conditions. A further increase in air will cause a clear stack; dense, white smoke indicates far too much air. Thus air must be regulated until a faint, light-brown haze issues from the stack.

Smoke can also be caused from:
1. Sprayer plates of unequal size in the same furnace.
2. Oil spray on the diffuser, carbon formation on the throat tile, a dirty atomizer, improper atomizer position, a damaged diffuser, worn plates or nozzles, no sprayer plate in the atomizer (firemen often forget to put them back when cleaning burners), a loose nut holding the plate in position, too much throttling of the air-register doors. If carbon forms on the throat tile and cannot be punched off without damaging the throat, remove the atomizer, crack the air-register doors, and let it burn off.

3. The atomizer may be in too far, causing incomplete mixing of fuel and air, and flame fluttering. If the atomizer is out too far, carbon forms on the burner throat. Both conditions cause smoke.

4. Fuel viscosity may be too high (oil not hot enough).

5. Air register and burner throat may not be centered, or the double front may be too deep, resulting in air leaking into the throat without passing through the register doors.

6. Air distribution in the double front may be faulty; check air pressure at various points in this area.

Q How is smoke density measured?

A A microRingelmann chart, Fig. 13-19, is used for comparing the smoke's density when no instruments are available. A more accurate method is measuring the stack gas with an in-stack gas analyzer, Fig. 13-20.

Q Explain the microRingelmann method.

A Ringelmann charts were originally used. These have five grids, No. 5 being solid black. But they have to be positioned 50 ft from the viewer, making them awkward to use. *Power* magazine omitted the black grid, and developed the smaller four-grid microRingelmann chart, which can be read at arm's length. To use:

1. Hold chart at arm's length and view the smoke through the slot in the chart.

2. Make sure the light shining on the chart is the same as that shining on the smoke being examined. For best results, the sun should be behind the observer.

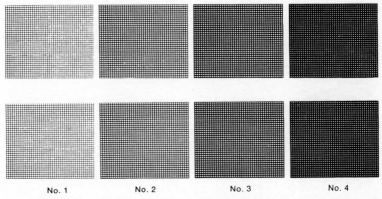

No. 1   No. 2   No. 3   No. 4

Fig. 13-19 MicroRingelman chart for measuring smoke density. These grids are a direct facsimile reduction of the standard Ringelmann chart as issued by the United States Bureau of Mines. (*Copyright* 1954 *by McGraw-Hill Publishing Company*, publisher of *Power*, 1221 *Avenue of Americas, New York, N.Y.* 10020.)

3. Match the shade of the smoke as closely as possible with the corresponding grid on the chart.

4. Enter the density of the smoke and the time of each observation on a record sheet.

5. Make repeated observations at regular intervals of ¼ or ½ minute.

6. To compute the smoke's density, use the formula: equivalent units of No. 1 smoke × 0.20 divided by number of observations = percentage of smoke density.

7. Note and record the distance to the stack, the direction of the stack (from where you are standing), the shape and diameter of the stack, and the speed and direction of the wind.

Most major cities have codes regulating the permissible smoke density, so be sure to get this information before fines are imposed on offenders.

Q Describe the in-stack continuous gas analyzer.
A There are various designs, some combining continuous monitoring with instant indication to allow fast response to excessive emissions by alerting the boiler-plant operator to air-pollution violations. Photoelectric detection is the principal measuring technique, with laser beams used in designs, Fig. 13-20, to minimize the effects of light scattering.

The basic minitoring system consists of a lamp and a receiving unit, mounted on opposite sides of the exhaust stack, as the figure shows. The sealed light source projects a beam of energy through particles of fly ash and unburned carbon traveling upward in the stack. The photocell receiver detects energy radiated from the light source and produces a minute electrical signal proportional to the energy detected. Since part of the radiant light energy is deflected or absorbed by suspended particulates, radiant energy detected by the photocell varies with the amount of suspended particulate matter.

The electrical signal is built up through high-gain amplifiers and is transmitted to a continuous-reading recorder, usually located at a central

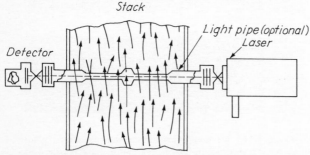

FIG. 13-20   In-stack continuous particulate emission detector.

control panel. In most systems, this recorder contains an alarm contact that operates when density exceeds a preset value. This contact can be wired to an external annunciator, signal light, or other type of alarm system.

BURNING GAS

Q Burning gas as fuel appears simple. Is it?
A *No.* It is really difficult and more dangerous than burning other fuels. Here are the reasons:

1. Low flame luminosity from many types of gas makes it hard to see what's going on in the furnace.

2. Invisibility of unburned gas accumulations presents an explosion hazard.

3. Varying heat content, ranging from 80 Btu per cu ft for blast-furnace gas, to about 1,000 Btu per cu ft for natural gas, to 2,000 Btu per cu ft for rich refinery gas.

4. Varying characteristics of flame-propagation speed and ignition temperature.

The ring burner is the most common type. Gas issues from a number of small holes around the inner periphery of the ring, jetting into a stream of air passing through the center. Damper vanes regulate quantity of entering air, usually giving it a spin to promote turbulence.

CAUTION: Always purge supply headers through a vent line to the roof before attempting to light off the first burner. Leave vent open until first burner is lit because the gas flow out of the vent makes it easier to control pressure in the burner header.

LIGHTING OFF THE FURNACE SAFELY

Q How should stoker-fed coal burning boilers be lit off?
A Run the stoker until the grates are partially covered with coal. Then put a quantity of light wood and some oily rags or paper on top and ignite under a light draft. As the coal catches on fire, increase feed and draft to produce the desired burning rate. Heat output is largely a matter of the amount of air supplied. Regulate coal feed to maintain a fuel bed thick enough to prevent blowing holes in the fire, yet thin enough to burn out by the time it reaches the discharge end of the stoker.

Q Explain steps needed to light off an oil burner safely.
A 1. Make sure oil is at the right temperature and pressure for the burner.

2. Make sure the damper is wide open. It should be so weighted that if holding screws let go, it will swing open.

3. Purge the furnace for a few minutes before applying the torch. If a blower is used, start it before oil is turned on.

4. Make sure that the oil burner is clean and in working order. If in doubt, insert a clean burner.

5. Insert a lighted torch. Place in the direct path of the oil, then turn the oil on.

6. Regulate the draft so the flame of the torch will not blow out. Keep the torch in front of the burner until the oil flame will maintain itself.

7. As the furnace temperature comes up, check burner and oil flame. Adjust oil flow and draft as necessary.

TWO IMPORTANT RULES: (1) Any increase in oil flow to burners *must be preceded* by increased air flow. (2) Any decrease in oil flow *must be followed by* decrease in air flow.

Q Should oil burners be fired off hot brickwork?

A *Never.* Numerous furnace explosions and personal injuries have resulted from this dangerous practice. *Always* use a torch.

Q Why, how, and how often are heating surfaces of a steaming boiler cleaned?

A On the fire side, soot and slag deposits not only reduce heat transfer rates from fuel to water, but buildup of such deposits increases draft loss and causes smoke. As the fire side of a boiler coats with soot, the exhaust stack temperature climbs, sending heat that could not be absorbed by the dirty tubes up into the atmosphere. Remember that $1/8$ in. of soot on a surface has the insulating effect of a $5/8$-in. layer of asbestos.

Steam or air is used to blow soot from heating surfaces, Fig. 13-21. Frequency of soot blowing depends on exhaust gas temperature rise, which tells the boiler operator when to blow tubes. When blowing, be sure to keep the boiler steaming at a reasonably high rate. This avoids the possibility of flameout or the explosion of dead pockets filled with un-

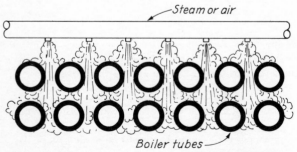

FIG. 13-21 Stationary soot blowers used for in-line tubes.

burned combustibles or fuel-rich gas in some portions of the boiler. Always increase draft slightly before blowing soot.

Water-washing the fireside is another method for keeping fire-side surfaces clean. This is done when the boiler is cold and out of service and the brickwork protected by tarpaulin (canvas).

SUGGESTED READING

REPRINTS SOLD BY *Power* MAGAZINE
 Coal, Selection and Handling, part 1, 24 pp.
 Coal, Combustion, part 2, 24 pp.
 Combustion, Pollution Control, part 1, 24 pp.
 Combustion, Pollution Control, part 2, 16 pp.

BOOKS
 Elonka, Stephen M., and Anthony L. Kohan: "Standard Boiler Operators' Questions and Answers," McGraw-Hill Book Company, New York, 1969.
 Elonka, Stephen M., and Alonzo R. Parsons: "Standard Instrumentation Questions and Answers," Vols. I and II, McGraw-Hill Book Company, New York, 1962.

14

BURNING COAL, OIL, AND GAS

Since the stoppage of fuel oil shipments from the Mideast to North America in 1973, many plants that had converted to oil or gas have turned back to burning coal. North America has abundant supplies of coal, which is stoker-fired for most industrial boilers. But only petroleum products can be used in internal combustion engines, for which they must be conserved.

Here we cover the methods and equipment needed for burning coal, oil, and gas, from which fuels the operator must squeeze the last possible Btu because of spiraling fuel prices and the restrictions on air pollution. And in case you ever have to hand-fire coal to at least supply heat for your plant, this is also covered.

HAND-FIRING COAL

Q What tools are used in hand-firing steam boilers?
A They are;

1. A *shovel* spreads fresh fuel on the fire and loads coal or ashes if they are handled by wheelbarrow.

2. A *slice bar,* before the fire is cleaned, breaks up clinkers that adhere to the grate. With the *alternate method* of cleaning fires, the bar sweeps fire from one side of the grate to the other.

3. A *hoe* pulls out ashes from the top of grate when the fire is being cleaned, and cleans out the ashpit.

4. A *rake* cleans fires and spreads fuel evenly over the grate.

5. A *lazy bar* is a short bar whose ends are hooked over the door hinge and latch of the fire door and ashpit door. It supports hoe and rake when they clean the fire and ashpit.

Q What hand-firing systems are used?

A 1. SPREADING METHOD: Coal is spread in an even layer over the entire grate at each firing, usually commencing at the back of the grate and working out toward the fire door. Coal must be fired frequently in small quantities.

2. ALTERNATE METHOD: Each side of the grate is fired alternately so that the volatile gases distilled from fresh fuel will be ignited by the bright fire on the other side of the grate.

3. COKING METHOD: Coal is fired at front of grate and allowed to coke there. Afterward it is pushed back and spread over the grates, and more coal is fired at the front. Thus the hot gases distilled from fresh fuel at the front of the grate ignite and burn as they pass over the glowing fire to the rear of the grate.

The alternate and coking methods are preferable to the spreading method when coal is high in volatile matter, since they do not lower the furnace temperature by blanketing the entire fire surface with fresh fuel at any one time. Also, there is less heat loss from volatile gases passing to the chimney unburned.

The spreading method, though probably most generally used, is efficient only when firing is light and frequent so as not to reduce the furnace temperature below the ignition point of the gases distilled from the fuel. Covering the entire fire surface at infrequent intervals with a thick layer of fresh fuel is most wasteful and inefficient.

Q Describe two methods of cleaning fires when hand-firing.

A Two methods are;

1. CLEANING ALTERNATE SIDES: The top layer of coked fuel on one side of the grate is swept over to the other side by the hoe and slice bar. The clinker and ash thus laid bare are pulled out with the hoe. The coked fuel is swept back over the clean part of the grate and fresh fuel fired. When the fire again burns brightly, the other side of the grate is cleaned in the same manner.

2. FRONT TO REAR: The coked fuel on the grate front is pushed to the rear after ashes and clinkers have been worked up to the front of the grate with the hoe. Ashes are then pulled out through the fire door and coked fuel pulled forward and spread evenly over the grate. The alternate is better than the front-to-rear method, but sometimes the latter is the only one that can be used where grates are small and fire door narrow.

Q Sketch and describe several grate bars for stationary grates.

A A *plain single-grate* bar is shown in Fig. 14-1*a*. This is a straight cast-iron bar with side projections at ends and center to keep the bars forming the grate far enough apart to leave an air space.

Figure 14-1*b* shows a *plain double-grate* bar and Fig. 14-1*c* a *herringbone-*, or *Tupper, grate* bar. The latter is wider and has less tendency to warp than the plain grate bar.

Figure 14-1*d* shows a grate bar for burning sawdust or other fine material. A large number of fine holes permit air for combustion to pass up through the grate, and yet allow very little unburned fuel to drop into the ashpit.

Q Sketch and describe a *pyramid-type* grate bar.

A This bar, Fig. 14-1*e*, has a layer of flat-topped hollow pyramids with slits in the sides through which air passes. The great number of small air openings gives very even air distribution through the fire bed, with no direct upward current and no large openings through which fuel falls into the ashpit. This grate is made in both stationary and shaking types.

Q How are shaking grates constructed, and what are their advantages over stationary grates?

A Shaking-grate bars are supported at both ends by pins resting on a notched side plate, and have projecting lugs attached to a long flat bar running the entire length of the grate, at the side. This bar extends through the boiler front or connects to an extension rod, Fig. 14-1*f*.

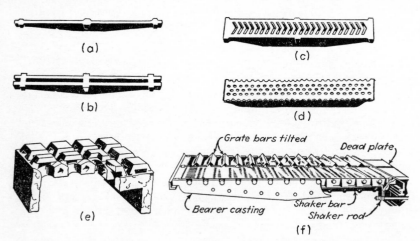

FIG. 14-1   (*a*) Plain single-grate bar; (*b*) plain double-grate bar; (*c*) herringbone-grate bar; (*d*) wood-burning-grate bar; (*e*) pyramid-grate bar; and (*f*) shaking-type grate.

Moving this rod to and fro by a hand lever or a power cylinder attached to the boiler front rocks or tilts all the grate bars. If the grate is made in two sections, two levers or rocking cylinders are used. Figure 14-1f shows a shaking grate with the bars on one side tilted, as they would be when fires are being cleaned, and the bars on the other side flat, as they would be in the usual firing position.

With stationary grates, fire doors must be kept open when the fires are being cleaned, but with shaking grates the fires can usually be cleaned by rocking the grate bars and working the ashes and clinkers down through the grate into the ashpit, without having to keep the fire doors open for long periods, or temporarily baring part of the grate and thus allowing excessive quantities of cold air to enter and cool the furnace.

Q What regulates size of air spaces in a coal-burning grate?

A The main factors regulating their size are the kind of coal being burned and the nature of the draft. Much coal is lost if a wide air space is used in grates burning slack or fine coal. Air openings can also be smaller if mechanical draft is used. In general, air spaces run from $1/4$ to $3/8$ in. for small coals and from $1/2$ to $3/4$ in. for larger sizes. The quantity of ash in the fuel and the coking or noncoking properties of the coal also have a bearing on the proper size of air space to use.

Q What causes smoke and how is its formation prevented?

A Visible smoke is composed of particles of soot formed by the incomplete combustion of the hydrocarbon constituents of the fuel. Smoke is produced if:

1. Air supply is insufficient for complete combustion.
2. Air and combustible gases are not thoroughly mixed.
3. Furnace temperature drops below the ignition point of the gases.
4. Combustion space is too small, preventing proper mixing of air and gases, and allowing gases to come into contact with cool boiler surfaces before they are ignited.

To prevent smoke formation, avoid these conditions. The air supply must be sufficient for complete combustion. Air and gases must mix thoroughly. Keep furnace temperature above the ignition temperature of the gases. There must be ample combustion space and proper baffling, where necessary, to mix air and gases and guide them on the proper path.

Q What provisions are made for preserving and extending the life of furnace walls in large boiler installations where combustion rates are very high?

A Where furnace temperatures are extremely high, use special air-cooled tiles, or protect the walls by banks of tubes through which water circulates. The latter is called a *waterwall*.

MECHANICAL STOKERS

Q Why are mechanical stokers preferable to hand firing?

A Less labor and attendance are needed to operate mechanical stokers. Cheaper grades of fuel are burned successfully. More fuel can be burned per square foot of grate surface. Much higher efficiencies can be attained. Better furnace conditions can be maintained. Production of smoke can be eliminated. Firing conditions can be controlled more exactly to meet varying loads, and overloads can be carried that would be impossible with hand firing.

Q Into what general classes may mechanical stokers be divided?

A Most stokers in common use fall into one or the other of these classifications: (1) sprinkler stoker, (2) traveling or chain-grate stoker, (3) overfeed stoker, (4) underfeed stoker.

1. SPRINKLER STOKER: May be used in conjunction with stationary grates, shaking grates, or even chain grates. In the last-mentioned case, there is no need to open fire doors or clean fires; good results are claimed for this combination. The sprinkler stoker spreads coal over the grate much the same as an expert hand fireman does with his shovel. It may be mounted on the boiler front above the fire doors, leaving them clear for cleaning fires if the grates are stationary or for hand firing if the stoker should break down or be cut out for repairs.

2. TRAVELING or CHAIN-GRATE STOKER: Coal is fed onto the front of a revolving grate, and burns as it travels toward rear of grate where the ashes are dumped over the back.

3. OVERFEED STOKER: Coal is fed onto a stepped reciprocating grate, either at the front or the side, and is gradually consumed as it travels toward the rear or the bottom of grate.

4. UNDERFEED STOKER: In one form, coal is fed into one or more deep retorts by power-driven rams and overflows onto side grates or dead

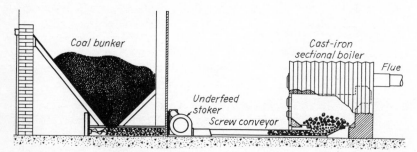

FIG. 14-2 Bin-fed stoker usually works well for a new smaller plant, or for a hand-fired boiler being modernized.

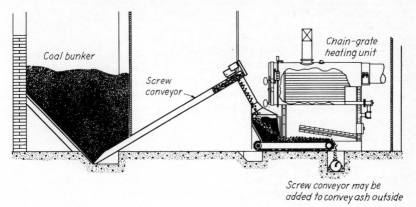

FIG. 14-3 Chain grate stoker can be fed by a screw conveyor from bin, and ash removed by another screw conveyor.

plates which slope downward to dump plates that can be tilted to dump ashes when necessary. In the other type, grates slope from front to rear, and coal is also fed by power-driven rams into narrow retorts between the grate bars, whence it flows over onto the grates. The grates have a reciprocating motion that moves the coal downward to dump plates at the bottom that discharge the ashes to an ashpit or hopper.

Chain-grate and overfeed stokers in small or medium-sized power plants may operate on natural, induced, or forced draft. Traveling or chain-grate stokers in large plants are usually constructed for forced-draft operation. Because of the thick fuel bed, the underfeed stoker is essentially a forced-draft stoker.

Q Is it practical to install stokers for burning coal when modernizing very small boiler plants?

A Yes, Fig. 14-2 shows one such installation for a cast-iron boiler plant. Here a screw conveyor is direct-fed from the bin. Similar systems are used in plants up to 100 boiler hp for such services as supplying heating steam for small apartment and office buildings, hotels, or for process steam on a year-round basis in smaller industrial plants, dairies, etc.

Q Are chain-grate stokers practical for smaller boiler plants?

A Figure 14-3 shows a suggested layout. This chain grate has the advantage of being a partial package-type unit with simple ash-removal arrangement. Ash is removed as it forms with a screw conveyor shown installed in a pit under the conveyor's end.

The amount of ash to be removed from a 90-hp plant, for example, may be about 15 tons per year in a lightly loaded heating plant. But it may be high as 65 tons per year in high-pressure year-round plants. Thus, with today's higher labor costs, mechanical devices for saving labor must be considered.

Q Describe a sprinkler stoker.

A Figure 14-4 shows a forced-draft sprinkler stoker with power-operated shaking grates. Firing doors are fitted to the boiler front beneath the stoker so that the boiler may be hand-fired if the stoker is stopped for any reason. When the stoker operates, a reciprocating feed plate feeds coal from the hopper into a compartment containing a revolving shaft with metal fingers that sprinkle coal over the fire with a spreading action. Rate of feed and sprinkler shaft speed can be adjusted to suit the boiler load. In some types of sprinkler stoker, parts that are exposed to excessive heat are water-cooled.

Q Describe construction and operation of traveling-grate and chain-grate stokers.

A In the *traveling-grate stoker* the small castings forming the grate surface are fastened to heavy transverse bars that are in turn attached to sprocket-driven chains. In the *chain-grate*, Fig. 14-5, the small grate castings are linked together by transverse rods, and the grate is driven by sprocket wheels engaging special sprocket links. Both grates form a continuous chain or band, and the process of combustion is the same in either case. Coal from a hopper is fed onto the front part of the grate through an adjustable door which regulates the thickness of the fuel bed.

The speed at which the grate travels should be regulated so that the coal is entirely burned to ash when it reaches the rear of the grate. The ashes fall over the rear of the grate into an ashpit or ash hopper which is

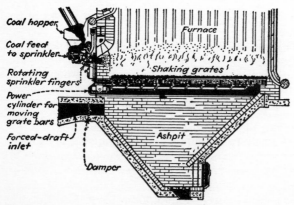

FIG. 14-4 Sprinkler stoker has shaking grates.

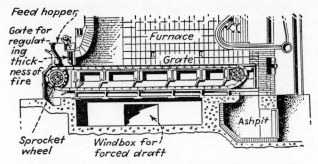

FIG. 14-5 Chain-grate stoker over forced draft.

usually emptied at intervals by a conveyor. With induced draft, the ashpit is open and air passes up freely through the grates. With forced draft, the ashpit is closed and the space inside the grate is divided into compartments or zones, thus enabling the air supply to be regulated to different parts of the grate to meet combustion needs. A side view of a forced-draft chain-grate stoker is shown in Fig. 14-5.

Q Sketch and describe a front-feed overfeed mechanical stoker.

A Figure 14-6 shows a side view. In this stoker, coal flows by gravity from a hopper onto a dead plate and is fed from there to the grate by a

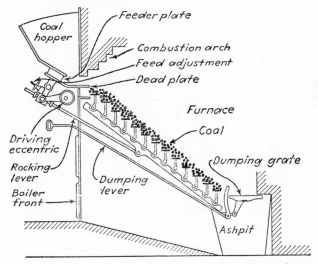

FIG. 14-6 Overfed inclined-grate stoker is slanted steeply toward the ashpit.

reciprocating pusher. The grate is steeply inclined, and each separate bar has a rocking motion that gradually works the coal down to the bottom of the grate. Ashes pass onto a dumping grate and are dumped at intervals into an ash hopper. The grate bars and pusher-feed plate are operated by an eccentric, and the eccentric shaft is driven by an electric motor or small steam engine or steam turbine.

Q Describe the operation of an underfeed stoker.
A There are a great many different types of underfeed stokers, but practically all operate on the principle illustrated in Fig. 14-7, which shows a sectional view of a forced-draft single-retort underfeed stoker. Coal from a feed hopper is pushed forward into a deep central retort by a power-driven ram and distributed the full length of the retort by auxiliary pushers. The coal spills over the sides of the retort onto inclined side grates, some of which have a reciprocating motion that feeds the coal slowly downward onto dump plates. These are tilted at intervals to deposit ashes in ashpits or ash hoppers. Ashes are removed through doors in the stoker front or, if hoppers are used, discharged into ash cars or conveyors.

The grate bars are hollow, and some of the air for combustion is admitted to the fire through openings in the grate bars at the point of distillation of the gases, that is, just above the fresh green fuel being fed in at the bottom of the retort. The rest of the air passes through the hollow grate bars to auxiliary wind boxes at the lower ends of the grates and is distributed from the boxes up through the fire bars to the fuel on the

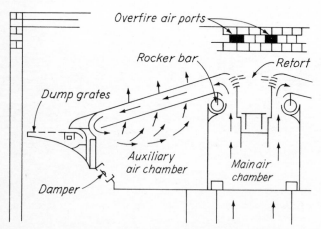

FIG. 14-7 Single-retort stoker is one of many different underfeed types in use.

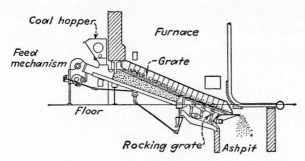

FIG. 14-8 Multiple-retort underfeed stoker has sloping main grate from front to rear.

grates. Air may also be admitted over the fire to assist combustion and prevent smoke.

Figure 14-8 is a side view of a multiple-retort underfeed stoker. It has a number of narrow retorts with narrow grates between. The main grates slope quite steeply from front to rear, and the coal feeds downward to a short grate with reciprocating bars that carry the ashes onto a dump plate. It is tilted periodically by hand or power and the ashes deposited in an ashpit or hopper. As in the center-retort stoker, power-operated rams feed the coal into retorts from beneath the grate, and forced draft distributes the air for combustion through the fuel bed.

BURNING PULVERIZED FUEL

Q How is pulverized fuel burned in boiler furnaces?

A The coal is ground to fine dust in a pulverizing mill. An air current blows this dust into the furnace where it burns almost like a gas. Figure 14-9 shows one type of pulverized fuel burner. The fuel-laden air current enters an outer chamber which is shaped to give the current a circular motion. This tends to throw the coal outward, so vanes are placed around the inner circumference to divert the stream toward the center and keep the coal-and-air mixture evenly distributed across the entire area of the pipe. The coal spreader at the burner tip breaks up the mixture into alternate lean and rich layers and causes the stream to assume a conical form.

Secondary air, supplied by a separate fan, is admitted around the tip of the fuel burner. The volume of secondary air can be regulated by dampers, and adjustable vanes give it a whirling motion before it enters the furnace. Mingling of the rotating fuel-laden air stream from the burner with the rotating current of secondary air tends to ensure thorough mixing of the air and fuel and thus complete combustion.

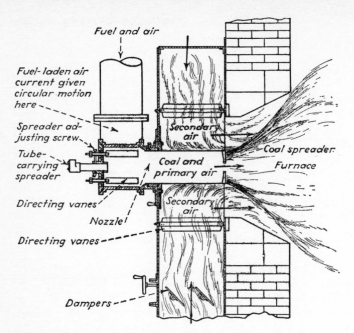

FIG. 14-9 Pulverized-coal burner spreads fuel into furnace.

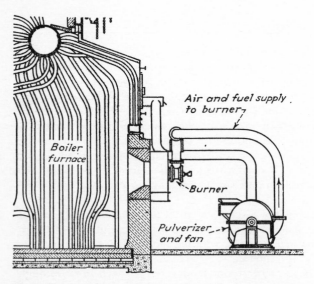

FIG. 14-10 Complete pulverized-coal-burning layout.

The fire is started by inserting an oil or gas torch through the tube in the front of the burner and then turning on the primary air and coal supply. Or a torch made from waste dipped in oil may be inserted through an observation opening. The torch should be in a position to ignite the fuel at the instant it is admitted to the furnace. If for any reason the fuel does not ignite, the torch should be withdrawn immediately and the setting allowed to clear before trying again to light up.

The secondary air supply is adjusted after the fire is lit. The shape or angularity of the burning flare can be changed by advancing or withdrawing the coal spreader by means of the tube in the center of the burner. Firing can be switched over to oil by inserting an oil burner in the center pipe.

Q Describe an arrangement for burning pulverized fuel in a boiler furnace.

A In Fig. 14-10, coal from a storage bunker feeds by gravity to a pulverizing mill. A fan draws the fine coal dust from the pulverizer and delivers it to the burners. Another fan supplies the secondary air needed for complete combustion.

BURNING GAS

Q Explain the construction and operation of low- and high-pressure gas burners.

A The important thing in all gas-burning devices is a correct air-and-gas mixture at the burner tip. Low-pressure burners, using gas at a pressure less than 2 psi, are usually of the multijet type, in which gas from a manifold is supplied to a number of small single jets, or circular rows of small jets, centered in or discharging around the inner circumference of circular air openings in a block of some heat-resisting material. The whole is encased in a rectangular cast-iron box, built into the boiler setting and having louver doors in front to regulate the air supply. Draft may be natural, induced, or forced.

Figure 14-11a shows a high-pressure gas mixer in which the energy of the gas jet draws air into the mixing chamber and delivers a correctly proportioned mixture to the burner. When the regulating valve is opened, gas flows through a small nozzle into a venturi tube (a tube with a contracted section). Entrainment of air with high-velocity gas in the narrow venturi section draws air in through large openings in the end. The gas-air mixture is piped to a burner. The gas-burner tip may be in a variety of forms. The one in Fig. 14-11b is called a sealed-in tip because the proper gas-air mixture is piped to the burner, and no additional air is drawn in around the burner tip. Size of the air openings in the venturi

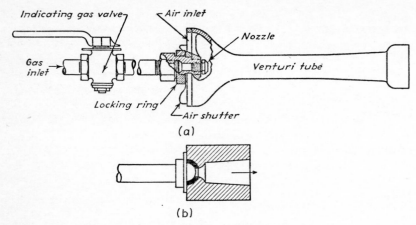

FIG. 14-11   (*a*) High-pressure gas burner; (*b*) sealed-in burner tip.

tube end is increased or decreased by turning a revolving shutter, which can be locked in any desired position.

BURNING FUEL OIL

Q Describe some type of steam-atomizing burner and explain its operation.

A Figure 14-12 shows the tip of a steam-atomizing oil burner with part cut away to show the internal construction. Figure 14-13 is a sectional view of the complete burner installed in a boiler setting.

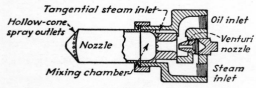

FIG. 14-12 Steam-atomizing burner tip.

The fuel oil entering the burner tip meets a high-velocity steam jet which forces it past a row of smaller steam jets into a mixing chamber. These small steam jets give the oil a whirling motion that tends to break it up into very fine particles through the action of centrifugal force. The atomized oil is then blown through small holes in the tip of the burner in the form of a cone-shaped spray. Air for combustion, controlled by adjustable louvers, passes through the burner. The burner and oil line can be blown clean with steam when the burner is not in use.

Q What is the first step in atomizing oil?

A First step in atomizing oil is heating it enough to get the desired viscosity. This temperature varies slightly for each grade oil. Lighter oil, numbered 1, 2, and 3, usually needs no preheating. The commercial grades 5 and 6 give best combustion results when heated enough to introduce the oil to the atomizer tip between 150 and 200 SSU (Seconds, Saybolt Universal viscosity). Then, after oil is heated to the right temperature, it must be pumped into the burner at the right pressure.

Q Describe a burner in which the oil is atomized by pressure alone.

A This type is usually called a *mechanical* burner to distinguish it from the steam-atomizing burner. The mechanical burner breaks up the oil into very fine particles by forcing it under high pressure through a number of very small passages in the tip. Oil pressures run from 30 to 300 psi, depending on the burner, and the oil must be heated to temperatures ranging from 100 to 300°F so that it will flow freely. Air for combustion is usually admitted through the burner casing, and the burners are blown out with compressed air when not in use, to prevent the oil from hardening and clogging the passages.

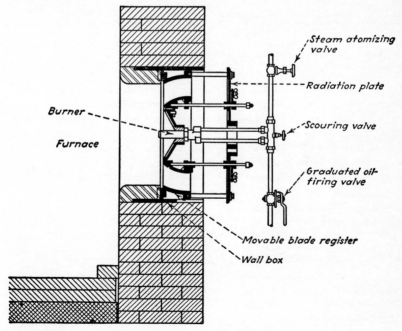

FIG. 14-13 Steam-atomizing oil burner with hand control valves.

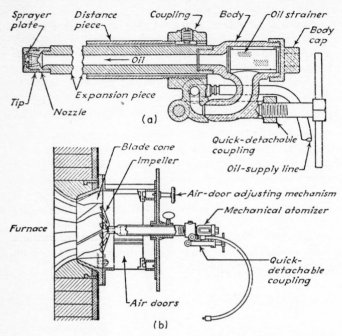

FIG. 14-14 (*a*) Mechanical oil-burner tip, (*b*) complete mechanical oil-burner installation.

A mechanical oil-burner tip is shown in section, Fig. 14-14*a*. In this burner, oil is forced through tangential slots in the nozzle against a sprayer plate; this causes the oil spray to issue from the burner in the form of a hollow cone.

Figure 14-14*b* shows the complete burner installation. Air for combustion enters through adjustable louvers which give it a rotary motion. This motion is further accentuated by the diffuser, or impeller plate, at the tip of the burner. The diffuser plate also splits up the air stream so that air and oil spray mix thoroughly before burning. Burners of the types in Figs. 14-13 and 14-14*b* are also made to burn either oil or gas. There are even triple-service burners that burn oil, gas, or pulverized fuel.

Q Sketch and describe a burner that burns oil or gas as required.
A A combination oil and gas burner is shown in Fig. 14-15. The oil-burning part is of the mechanical type, somewhat like that of Fig. 14-14. The gas burner is a ring with many small jets on the inside. The air supply is controlled by opening or closing the movable blade registers.

Q Sketch a method of installing oil or gas burners in a boiler furnace.

A Because the burning characteristics of oil and gas are much alike, the same method of installation does for both, and very often combination burners are used that burn either oil or gas. Figure 13-13 in the previous chapter shows oil burners under a water-tube boiler.

Loose firebrick checker walls and firebrick baffle walls are sometimes placed in gas- or oil-fired furnaces to assist in the mixing and ignition of the gas-air mixture. When using these, take great care to construct them so that heat is not concentrated on some particular part of the boiler, thus burning or bagging tubes or plates.

Q What precautions should you observe when firing a boiler with gas?
A 1. See that all gas valves are tight and that there are no leaks in pipes or fittings.

2. Perforate all dampers to ensure sufficient draft, even when dampers are closed, to carry off any gas that may enter the furnace through a leaky valve or from the accidental opening of a gas valve.

3. Before lighting burners, make sure gas valves are tightly closed, dampers are open, and there is no trace of gas already in the furnace. Use a proper torch for lighting up.

4. If burners blow out from any cause, allow a long enough interval to clear the furnace of gas before lighting up again. Do not attempt to light

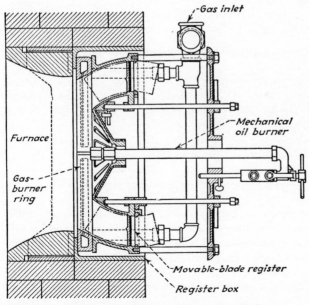

FIG. 14-15 Combination oil and gas burner installation.

burners from hot brickwork; use a torch. Many serious accidents have occurred from neglecting these precautions.   '

5. When starting to fire up, warm furnace slowly so as not to damage the setting or the boiler through excessive expansion of some parts while others remain cold.

6. When forced or induced draft is used, start the draft fan and let it run for a short time before lighting up in order to sweep away any possible accumulation of gas in the furnace. Keep the fan running for some time after shutting off the gas supply, to clear the furnace of unburned gas so there will be no danger of explosion.

Q What precautions should you observe when firing a boiler with oil?
A 1. See that oil pressure and temperature are correct before lighting up and, when steam-atomizing burners are used, that there is sufficient steam pressure. Open dampers.

2. If the oil spray does not ignite at once, or if the burners blow out from any cause, allow enough time to elapse to clear the furnace of oil vapor before lighting up again.

3. If for any reason unburned oil collects on the furnace floor, the gas distilled from this oil must be carried away and the furnace entirely cleared of any explosive gas-and-air mixture before attempts are made to light the burners.

1. Use strainers where necessary in oil lines and clean them regularly.

2. When atomizers are not in use, remove and thoroughly clean them.

Q What are the characteristic effects attending the burning of anthracite, bituminous, and lignite coal?
A *Anthracite,* being mainly composed of fixed carbon and ash, is low in gaseous constituents. Hence it ignites slowly, burns with a bluish flame, and gives off little smoke. For best results with hand firing, don't slice or disturb the fuel bed more than necessary.

Bituminous is more widely used than anthracite for firing power boilers. It varies greatly in composition but always has a higher volatile content than anthracite. In general it burns with a yellow flame. It may be coking or noncoking. Ash content and moisture may vary from very low to very high.

Lignite is usually lower in carbon content than bituminous and fairly high in volatile matter. Its moisture content is notably high. It burns with a smoky yellow flame. Though much lower in heating value than anthracite or bituminous coal, it is a satisfactory fuel when burned on grates and in furnaces of suitable design. Its high moisture content makes lignite hard to store for any great length of time without considerable slacking.

Q What are the advantages of oil and gas fuel compared with coal?

A   Apart from cost, which varies with the locality, oil and gas possess considerable advantage over coal as boiler fuels. Oil requires less storage space, leaves no ash, and burns better because of more intimate mixing with the air. Furnace and boiler room can be kept cleaner, labor costs lower. Very little excess air is needed to ensure complete combustion. Gas possesses all the advantages of oil, plus elimination of storage space and of steam or mechanical atomizing requirements.

Both oil and gas are smokeless fuels if furnaces and burners are properly designed. Standby losses are small because the fuel supply can be instantly shut off or reduced to a minimum when the load is off.

Q   Why, then, is coal the primary boiler fuel in North America?
A   Because coal is far more plentiful than oil or gas and is generally available at a sufficiently lower cost on a Btu basis to offset the listed advantages of oil and gas in most regions.

Q   What are the advantages and disadvantages of pulverized fuel compared with stoker-fired solid fuel?
A   In general, with pulverized coal it is easier to get proper mixing of fuel and air and to reduce excess air percentages and standby losses. Also, with suitable furnaces, there may be greater flexibility in coal choice.

Disadvantages with certain coals may include slagging of tubes and furnace walls. Pulverized firing adds the cost of pulverizing, and usually discharges more ash from the stack.

Q   What are the main requirements for an efficient boiler furnace?
A   It should have enough volume to contain all the gases that may be developed at overratings, and allow them to mix thoroughly with enough air for complete combustion. To ensure complete mixing and ignition, it should have a long flame travel and a tortuous path for gas travel, using baffle walls and ignition arches where necessary.

Q   What are the main reasons for low efficiency in boiler furnaces?
A   Low efficiency may be caused by poor design and bad workmanship. Combustion space may be too small, gas travel too short, baffles in the wrong place, refractory walls not properly bonded to outer setting.

Lowered efficiency in existing furnaces may be caused by excessive leakage of cold air into furnace through leaky brickwork, and also by poor heat transmission through accumulation of soot or slag on boiler-heating surfaces.

Q   How is air for combustion preheated, and what are the advantages of preheating?
A   Air for combustion purposes may be preheated before it enters the

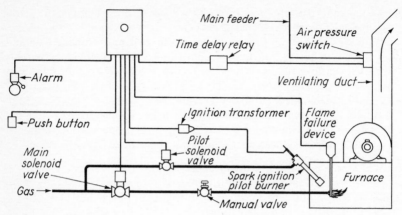

FIG. 14-16 Purging protection for gas-burning furnace is extremely important.

boiler furnace by passing it through banks of tubes placed in the flue leading from the boiler to the chimney, and thus using some of the heat that would otherwise pass to waste. Preheating has many advantages. Combustion is improved, efficiency and capacity are increased, and lower-grade fuels can be burned successfully. The hotter air supply increases the furnace temperature, with a consequent increase in combustion efficiency and greater transfer of heat to the boiler-heating surfaces.

Q Why should a gas-burning furnace have an air-pressure switch hooked into the system as shown in Fig. 14-16?
A The air-pressure switch in the ventilating duct in Fig. 14-16 prevents ignition of accumulated gas if the furnace is not properly purged before it can be relighted. The flame failure device shuts off the gas if the flame fails unexpectedly. To prevent furnace explosions, every gas-burning furnace must have a safety hookup for proper purging.

As an added safety precaution, every gas-fired boiler should be checked yearly by the gas supplier or contractor for proper operation and analysis of the flue gas for proper CO_2 content. This is not only a safety precaution, but also is necessary for energy conservation.

SUGGESTED READING

REPRINTS SOLD BY *Power* MAGAZINE
 Power from Coal, parts 1, 2, and 3, each 24 pp.
 Steam Generation, 48 pp.
 Fans, 32 pp.

BOOKS

"North American Combustion Handbook," North America Manufacturing Co., Cleveland, Ohio, 1972.

Elonka, Stephen M., and Anthony L. Kohan: "Standard Boiler Operators' Questions and Answers," McGraw-Hill Book Company, New York, 1969.

Elonka, Stephen M.: "Standard Plant Operators' Manual," McGraw-Hill Book Company, New York, 1975.

15

FEED-WATER TREATMENT

Natural waters contain impurities, depending largely on the source. Wells and springs are classed as ground water, rivers and lakes as surface water. Ground water picks up impurities as it seeps through the rock strata, dissolving some part of almost everything it contacts. But the natural filtering effect of rock and sand usually keeps it free of suspended matter.

Surface water often contains organic matter such as leaf mold and insoluble matter such as sand and silt. Pollution from industrial waste and sewage is frequently present. Stream velocity and amount and location of rainfall can rapidly change the character of the water.

In general, impure water tends to cause trouble in industry by forming deposits on surfaces, thus impairing heat transfer (as from fire side to water side of boiler) or product quality. Also, it may damage metal, wood, or other construction materials used to store, transport, or otherwise contain water. Thus, the broad aims of treatment are to prevent deposits and control corrosion. And this must be done at a cost that is economically justified by benefits accrued from use of trouble-free water.

For industry, undesirable water impurities fall into these main groupings: (1) dissolved mineral and organic matter, (2) dissolved gases, such as oxygen and carbon dioxide, (3) turbidity, (4) color, (5) taste and odor, and (6) microorganisms.

IMPURE FEEDWATER

Q What are the effects of impure feedwater in boilers?

A Effects are (1) foaming and priming; (2) mud deposits on heating surfaces with consequent danger of burning or bagging plates; (3) scale deposits on heating surfaces which retard transmission of heat through metal, thus wasting heat and causing danger of plates' overheating and burning or bagging; (4) corrosion of plates and other metal surfaces.

Q What are the common impurities in boiler feedwater?

A Impurities in boiler feedwater may be divided into:

1. SCALE-FORMING SUBSTANCES: The principal ones are the carbonates and sulfates of lime and magnesium.

2. SCUM-FORMING SUBSTANCES (may be mineral or organic): (*a*) Mineral scum-forming or foaming impurities usually contain soda in the form of a carbonate, chloride, or sulfate. (*b*) Organic impurities that form scum or cause foaming are usually found in water containing sewage.

3. SLUDGE-FORMING SUBSTANCES: There are usually solid mineral or organic particles, carried in suspension.

4. CORROSIVE SUBSTANCES: These may be chemical compounds, chloride of magnesium, free acids, or gases such as oxygen and carbon dioxide.

Q What are the chemical impurities commonly found in boiler feedwaters?

A Principal impurities, with their technical names, chemical symbols, common names, and effects, are given in the table below.

Name	Symbol	Common name	Effect
Calcium carbonate	$CaCO_3$	Chalk, limestone	Soft scale
Calcium bicarbonate	$Ca(HCO_3)_2$		Soft scale
Calcium sulfate	$CaSO_4$	Gypsum, plaster of Paris	Hard scale
Magnesium carbonate	$MgCO_3$	Magnesite	Soft scale
Magnesium sulfate	$MgSO_4$	Epsom salts	Corrosion
Silicon dioxide	SiO_2	Silica	Hard scale
Calcium chloride	$CaCl_2$		Corrosion
Magnesium bicarbonate	$Mg(HCO_3)_2$		Scale, corrosion
Magnesium chloride	$MgCl_2$		Corrosion
Sodium chloride	$NaCl$	Common salt	Electrolysis
Sodium carbonate	Na_2CO_3	Washing soda or soda ash	Alkalinity
Sodium bicarbonate	$NaHCO_3$	Baking soda	Priming, foaming
Sodium hydroxide	$NaOH$	Caustic soda	Alkalinity embrittlement
Sodium sulfate	Na_2SO_4	Glauber's salts	Alkalinity

Q What is meant by *hard* and *soft* water?

A *Hard* water contains an excess of scale-forming impurities, while *soft* water contains little or no scale-forming substances. Hardness is mainly a result of the presence of the carbonates of calcium and magnesium and can easily be recognized by its effect on soap. Much more soap is required to make a lather with hard water.

Q How is hardness expressed?

A In United States practice, it is expressed either in parts of calcium carbonate per million parts of water (ppm) or in grains of calcium carbonate per gallon.

Q What is the relation between ppm and grains per gallon?

A 1 ppm = 0.058 grains per U.S. gal
  1 grain per U.S. gal = 17.1 *ppm*

Q What are the principal scale-forming impurities in boiler feedwater, and what kind of scale does each deposit?

A Principal scale-forming impurities are the sulfates and chlorides of lime and magnesium, which form a hard scale, and the carbonates of lime and magnesium, which form a soft scale. Silica also forms a dense hard scale and has a hardening influence on the soft-scale-forming substances.

Q Explain what is meant by *temporary* and *permanent* hardness and the treatment necessary to reduce each.

A Water containing only the carbonates of lime and magnesium is said to have *temporary* or *carbonate* hardness because these impurities can be caused to precipitate as a soft sludge if water is heated to a temperature close to boiling point at atmospheric pressure.

Water containing the sulfates of lime and magnesium is said to have *permanent* or *noncarbonate* hardness because these impurities do not precipitate as solids until the water is heated to a temperature over 300°F. Thus, where water passes directly through an open heater to the boiler, heater temperature is not high enough to precipitate the impurities causing permanent hardness, and they therefore deposit in the boiler.

PURIFYING WATER

Q What methods are commonly used to purify feedwater or neutralize the bad effects of impure feedwater?

A There are three general methods of treatment: (1) mechanical, using filters and settling tanks; (2) chemical, using chemical treatment before or after feeding the water into the boiler; (3) thermal, using feed-water heaters.

Q What are the mechanical methods of feed-water treatment?
A Water containing solid matter in suspension may be purified (1) by chemical means alone; (2) by settling tanks or filters; or (3) by both. Settling tanks are large tanks into which the feedwater is pumped or run by gravity and allowed to stand for a time before being pumped into the boiler. This permits most of the suspended solids to settle to the bottom of the tank whence they are washed out through a drain at intervals.

Filters are tanks containing charcoal, coke, excelsior, or other finely divided filling through which the feedwater passes before it reaches the boiler feed pumps.

Q What are the chemical methods of treating boiler feedwater?
A There are two general methods:

1. Water is pretreated in a suitable purifying apparatus by adding chemicals to neutralize any acids present and precipitate the scale-forming impurities. The water is then filtered to remove these solids before it passes into the boiler.

2. Chemicals are simply added to the water as it is pumped into the boiler, or they are introduced separately into the boiler. Thus the solids precipitate and accumulate in the boiler itself.

Outside treatment is generally much better. With internal treatment, though the impurities are neutralized and rendered nonscaling and non-corrosive, there is a heavy concentration of chemicals and possible solids in the boiler itself.

Q What should be done before adopting any form of chemical treatment for boiler feedwater?
A Before starting to use any treatment, send samples of the feedwater to a competent analyst, who will report on the nature of the impurities and recommend the proper chemical treatment. Follow the analyst's advice. See Fig. 20-2 in Chap. 20.

Q Explain *coagulation*.
A This is partly a mechanical and partly a chemical process of water purification used in connection with filtering. Much of the impurity in surface waters is in a finely divided state and takes a long time to settle, or it may even be fine enough to pass through voids in the filtering material. By adding certain chemicals to the water, gelatinous or jellied substances are formed which cause the small particles of mud and silt to coalesce into groups too large to pass through the filtering material.

Q What are boiler compounds?
A They are commercial preparations in powdered, solid, or liquid form, made to treat impure boiler feed water. Carbonate of soda, caustic soda, phosphate of soda, and tannin are common ingredients of boiler com-

pounds. It is possible, though unlikely, that a compound may prove quite beneficial even if no tests have been made to determine its suitability, but the blind buying of a chemical compound guaranteed to cure all boiler ills is a mistake. No single mixture of chemicals can possibly be effective in neutralizing the impurities in all feedwaters.

If you use a boiler compound, buy it from a reputable company that makes a specialty of analyzing feedwater and making up a compound to suit each case, or submit samples of the feedwater and the proposed compound to a qualified analyst, for test and advice on the compound's suitability for treating the feedwater. Further tests should be carried out at fairly close intervals, as the nature of the impurities in the feedwater may vary greatly from time to time.

Q What are the commonest chemicals used in feed-water treatment?
A Slaked lime [$Ca(OH)_2$] and soda ash ($NaCO_3$) are the commonest. Slaked lime reduces temporary or carbonate hardness caused by the carbonates of lime and magnesium. Soda ash reduces permanent or noncarbonate hardness caused by the sulfates of lime and magnesium.

Q Describe the lime-soda treatment for boiler feedwater.
A In this process lime is added to the water to reduce the carbonate hardness. The carbonates of calcium and magnesium, which are the main impurities producing temporary hardness, are usually combined with dissolved carbonic acid (CO_2), forming bicarbonates which are soluble in water. The lime combines with the carbonic acid, so the bicarbonates are then reduced to calcium carbonate and magnesium hydroxide, which are insoluble and precipitate as a sludge.

Noncarbonate, or permanent, hardness resulting from the sulfates of lime and magnesium is reduced by adding soda ash. This decomposes the soluble sulfates into carbonates, which precipitate as solids, and soluble salts, which do not form scale in the boiler.

The water can be treated cold, but the chemical reactions are much quicker if it is heated almost to the boiling point. Use settling tanks and filters to remove the sediment.

Most feed-water treatments do not produce "pure" water. They may really increase the impurities, as chemicals are added in most cases, but the original objectionable impurities are now changed to a less harmful form.

Q Describe the hot-process feed-water-treating plant illustrated in Fig. 15-1.
A In Fig. 15-1, raw water enters the top of an open heater where its temperature is raised by direct contact with exhaust steam. This heating eliminates some of the dissolved gases carried by the water, and vents them to

the atmosphere through the vent condenser. The treating chemicals are added to the water as it passes into the sedimentation tank beneath, and the sediment deposited by the combined heating and chemical action is washed out at intervals through a drain in the bottom of the tank. Treated water is drawn off through the inverted funnel and, if necessary, passed through a filter before it reaches the boiler feed pump. The water takes 1 hr or more to pass through the system.

Q Describe the zeolite process of softening water.

A This process uses a sandlike substance called *zeolite* (natural or artificial) arranged in a filter bed. This substance has the remarkable property of *base exchange.* When hard water passes through a bed of zeolite, the calcium and magnesium in the hardness compounds pass *into* the zeolite and are replaced by sodium *from* the zeolite. Thus calcium bicarbonate becomes sodium bicarbonate, and magnesium sulfate becomes sodium sulfate. Since none of the sodium compounds are scale-forming, this exchange frees the water from hardness.

As this process continues, the sodium supply of the zeolite bed is gradually depleted until it no longer properly softens the water. The zeolite is then restored to its original condition by the action of a strong brine of common salt. This puts sodium back into the zeolite and removes calcium and magnesium.

Zeolite-treated waters show *zero hardness* by the soap test but are, of course, laden with soluble sodium salts. To avoid foaming and priming,

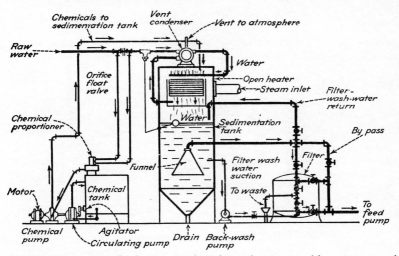

FIG. 15-1 Hot-process feed-water treating plant mixes steam with raw water and chemicals.

the concentration of these salts in the boiler must be kept within limits by suitable blowdown.

Because of the danger of embrittlement under certain conditions, a high percentage of sodium carbonate is undesirable in boiler water. Thus, zeolite softeners are better suited for waters in which sulfate hardness predominates. Sometimes zeolite softening is preceded by hot-process lime treatment to remove most of the carbonate hardness.

Figure 15-2 shows a modern zeolite softener with automatic control. During the softening part of the cycle, water flows downward through the zeolite bed. When the water-softening capacity of the zeolite approaches exhaustion, the automatic valves cut off the downward flow and backwash upward to loosen the material and remove deposited dirt. The third step is admission of a measured quantity of common salt brine at the top of the bed. After a suitable interval, a stream of rinsing water removes excess salt and cleans the zeolite, which is now ready for another softening cycle.

The unit pictured makes the various valve changes automatically at the right time, as determined by the water meters.

Q Explain the zeolite softener monitoring alarm system shown in Fig. 15-3.

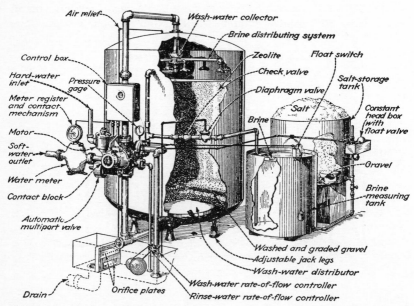

FIG. 15-2 Zeolite feed-water treating plant is old standby in boiler room.

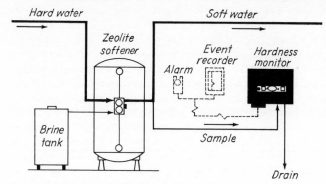

FIG. 15-3 Water-hardness monitor for zeolite softener.

A This water-hardness monitor keeps a close check on the water coming out of the zeolite softeners. As soon as the hardness rises to the crucial point, the monitor signals an alarm. The operator knows when softened water is flowing even with fluctuating hardness levels in the makeup water. It is one way to make sure that only soft water is going to the boilers.

Q What are the thermal (heating) methods of feed-water treatment? **A** In thermal purification the water is heated before it enters the boiler to a temperature where the chemical impurities in solution will solidify and deposit in the heater. Feed-water heaters are of two main types: *open* and *closed*. Open heaters are commonly used both to heat and to purify water containing carbonate hardness. Closed heaters also reduce noncarbonate hardness, but the difficulty of cleaning the tubes of a closed heater is a strong objection to its use as a water purifier.

The open heater is usually open to the atmosphere. The essential feature of an open heater is that it contains no heat-transfer surfaces, the steam and water being in direct contact. Usually open heaters operate at atmospheric pressure, or very slightly higher. Hence it is impossible to raise the temperature of the water in the heater much above the boiling point at atmospheric pressure (212°F at sea level). In practice, the temperature is often kept a few degrees below the boiling point because of the difficulty of pumping very hot water. The boiler feed pumps are placed between the open heater and the boiler so that they draw water from the heater and discharge it to the boilers. The pumps must be placed considerably lower than the heater so that the water will run into the pump suction by gravity. Water close to 212°F cannot be raised by pump suction. Open heaters are generally placed on a high platform in the boiler room.

A closed feed-water heater is simply a steel shell containing rows of tubes. Live or exhaust steam enters this shell and passes around the tubes (ordinarily). The feed water circulates through the tubes and is warmed by the heat conducted through the walls of the tubes from the steam. Since the closed heater is usually placed between the feed pump and the boiler, the water in the heater is always under boiler pressure, or higher, and its temperature thus can be raised much higher than that of the water in an open heater.

HEATERS

Q How is water purified in an open heater?
A Heating water almost to the boiling point causes the carbonates of lime and magnesium to solidify and deposit in the heater. These are the impurities that cause temporary, or carbonate, hardness. Water containing these impurities would deposit a soft porous scale in the boiler. Gases such as oxygen and carbon dioxide, which would cause corrosion in the boiler, are also driven off in an open heater and vented to the atmosphere.

Open heaters usually contain filtering compartments to retain the precipitated impurities, and also any other solid matter that may be in the feedwater.

Q What is the action in a closed heater?
A It is sometimes possible to raise the temperature high enough in a closed heater to cause the most troublesome impurities, the sulfates of lime and magnesium, to solidify and precipitate, as well as the carbonates of lime and magnesium which deposit at a much lower temperature. Sulfates remain in solution up to 300°F, at which temperature they solidify and settle. Because these impurities may form a hard scale on the heater tubes, it is not particularly profitable to use a closed heater as a water purifier. Water containing these impurities should be chemically treated before it passes through the heater.

Q Describe the open feed-water heater shown in Fig. 15-4.
A This is a common type of direct-contact open heater, using exhaust steam to heat the feedwater. Entering steam passes the oil baffle on the side and mingles directly with the water, which enters at the top and trickles down over the trays. The purpose of the trays is to break the water up into a great number of fine streams and thus allow it to mix more intimately with the steam. The heated water passes through a coke filter before entering the pump suction. An overflow pipe is provided to prevent flooding of the heater. A float-controlled valve holds the water level constant.

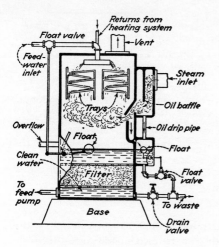

FIG. 15-4 Open feed-water heater.

The feed pump must be placed at a lower level than the heater, as the pressure in the pump suction is always less than atmospheric. If the temperature is carried too close to the boiling point in the open heater, the water will flash into steam in the suction line and pump cylinders, unless the heater level is many feet higher than pump suction.

Q Describe a closed feed-water heater.

A Figure 15-5 shows a common type. This heater consists of a steel shell containing a large number of small tubes through which water passes

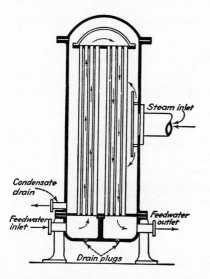

FIG. 15-5 Closed feed-water heater.

while steam circulates on the outside. The water and steam do not come into actual contact as they do in the open heater. Either exhaust or live steam can be used for heating the water, but the use of live steam for this purpose does not save any fuel.

The closed feed-water heater is usually placed between the feed pump and the boiler. Hence the water passing through the heater is under a pressure higher than the boiler pressure, and its temperature can be raised much higher than in an open heater. With this arrangement, water is heated after leaving the feed pump, so the pump does not have to handle the heated water.

Many plants use an open heater before the pump, and a closed heater (or several closed heaters in series) after the feed pump.

Q What is an *evaporator?*
A It is an apparatus for securing pure feedwater by distillation. A bent-tube evaporator, operating on much the same principle as a fire-tube boiler, but using steam instead of hot gas as the heating medium, is shown in Fig. 15-6. Steam enters at the top of the manifold, passes through the rows of bent tubes, returns to the bottom of the manifold, and drains away as condensate. Water in contact with the outer surfaces of the tubes evaporates and passes out as vapor at the top of the shell, either to a closed condenser, such as a closed feed-water heater, or to an open feed-water heater, where it is condensed.

Distillation of the water in this way removes dissolved gas, but all solid impurities are left in the evaporator, which must be cleaned regularly to remove scale from the outside surface of the tubes. One method of scale removal is to crack it off quickly by varying the steam and water tempera-

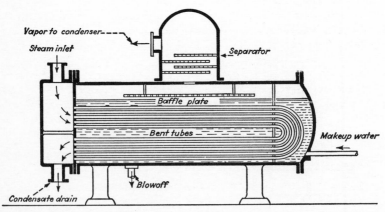

FIG. 15-6 Evaporator distills water heated by steam.

tures, causing the tubes to expand and contract. Sometimes a water softener is used ahead of the evaporator to prevent scale formation, in which case the impurities are deposited as a soft mud and can be removed by blowdown. Evaporators are commonly used in condensing power plants to purify the raw makeup water required for boiler feed. In most industrial plants the large percentage of makeup would make evaporators too large and costly.

OPERATING PROBLEMS

Q What are the cause and cure of *foaming* and *priming?*

A *Foaming* is said to occur when the steam space in a boiler is partially filled with unbroken steam bubbles. *Priming* is the carrying over of liquid water from the boiler by the steam.

Foaming is caused by impurities in the water preventing the free escape of steam as it rises to the surface, or by an oily scum on the surface of the water. This surface scum may be caused by oil, vegetable matter, or sewage in the water. The remedies are treatment and filtration of the water to remove solid impurities, plus enough blowdown to avoid excessive concentration of salts.

An exceedingly dangerous operating condition, foaming indicates a pressing need for boiler cleaning, feed-water treatment, or both. It is practically impossible to tell the true water level in a badly foaming boiler, and therefore it may be difficult to prevent damage due to low water.

Priming, carrying over of water slugs with the steam, may be caused by the same conditions that cause foaming, but it may occur with perfectly pure feedwater if the steam space is too small, if the boiler is forced way above its normal capacity, or if the water level is carried too high. Priming is highly dangerous because of the possibility of suddenly lowering the water level below the danger line, and also because the water slugs carried over may do great damage to pipe lines, valves, fittings, pump, and engine cylinders.

To avoid priming and probable water-hammer damage through sudden opening of large steam valves, all main stop valves on boilers and steam lines, and throttle valves on engines, should be opened very slowly and water in the boiler should not be carried above its normal level. In plants with super-heaters any liquid water going to the superheater is there evaporated, so troublesome deposits are formed in the superheater. Liquids carried over may damage steam turbines.

Q What are the causes of *corrosion,* and how do you prevent it?

A *Corrosion* is the wasting away of the metal surfaces of boiler parts. It may be external or internal. External corrosion is a rusting process caused by water leaks or drips on the outside surface of the boiler, or at places hidden by insulation or brickwork. It may also occur in fire tubes and flues from the action of sulfuric acid formed from moist soot and the sulfur dioxide in combustion gases. Internal corrosion is caused by acids, oxygen (from air) or other gases in the feedwater, or by electrolytic action.

Corrosion evenly distributed over the metal surface is called *uniform* corrosion; it may be in the form of little holes or pits, called *pitting*, or it may occur as the eating away of a narrow groove along the edge of a riveted joint, called *grooving*. The latter is induced by the slight bending action of the metal along the riveted joints as the boiler expands and contracts. This movement keeps a narrow strip of metal surface clean and free from scale, thus offering a clean surface for corrosive elements to attack.

External corrosion may be prevented by keeping external surfaces, also tubes and flues, clean and dry. Internal corrosion may be prevented by suitable water treatment and the removal of dissolved gases, particularly oxygen.

Q What is *caustic embrittlement,* and how do you prevent it?

A So-called *caustic embrittlement* is a hardening and crystallizing process, sometimes called *intercrystalline corrosion*, which causes fine cracks in metal, particularly in riveted joints and around rivet holes. Its exact cause has not been fully determined, but it commonly occurs where feedwater contains a large amount of free bicarbonate of soda or has been treated with excessive amounts of soda ash, caustic soda, or compounds containing these chemicals. This is one of the dangers of using a boiler compound without knowing its exact composition and the feedwater's impurities.

Sodium bicarbonate and sodium carbonate *hydrolyze* under heat and pressure into sodium hydroxide (caustic soda). This caustic soda tends to concentrate in the riveted joints where its contact with the steel is believed by some authorities to liberate free hydrogen gas, which penetrates and embrittles the steel.

Results of extensive tests in laboratories and in boilers under ordinary working conditions seem to indicate that embrittlement can be prevented if all alkalies that might yield sodium hydroxide are eliminated and if sodium phosphate is used to maintain the correct degree of alkalinity in the boiler. Tannin and lignin derivatives have proved to be effective inhibitors of embrittlement over a wide range of boiler pressures.

All-welded boilers offer no cracks for concentration of caustic soda and hence should be free from embrittlement trouble. The same holds true of boilers with riveted joints if they are calked inside and not outside, so that any parts that are not tightly calked can be checked by external signs of leakage through the joint.

Q For what are *deconcentrators* and *continuous blowdown* systems used?
A When there is high concentration of soluble or suspended solids in boiler feed water, some will undoubtedly precipitate as sludge and bake into scale on tubes and plates. This condition can and should be remedied by proper treatment and filtering to remove troublesome solids before the feedwater enters the boiler. Where further filtering is desirable, some of the boiler water can be removed continuously, passed through a deconcentrator, which is simply a pressure filter, and returned to the boiler.

Strong concentrations of soluble impurities, while they may not be scale-forming, may cause foaming, priming, and even corrosion. Blowing down the boilers at intervals will reduce this concentration. Figure 15-7 shows the principle of continuous blowdown. It keeps concentration within reasonable limits and at the same time prevents loss of heat if the blowdown water is passed through a *heat exchanger* where its heat is given up to incoming feedwater. The heat exchanger is similar to a closed feedwater heater or a shell-and-tube condenser. Hot blowdown water passes through the tubes contained within the shell, and the cold water circulates around the outside of the tubes, or vice versa. A control valve placed between boiler and heat exchanger controls blowdown flow rate.

Q What is meant by the pH value of boiler feedwater?
A All water contains free hydrogen (H^+ ions) and hydroxide (OH^- ions). H^+ ions give acid reactions, and OH^- ions give alkaline. In pure water that is neither acid nor alkaline these ions balance each other, and the water is

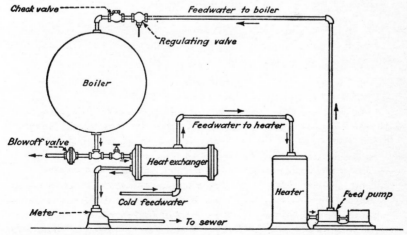

FIG. 15-7 Continuous blowdown system removes impurities.

neutral. If the balance is disturbed so that H^+ ions predominate, the water will be acid; if OH^- ions are in excess, it will be alkaline.

In feed-water treatment it is important to know the degree of acidity or alkalinity, and the pH system of notation has been devised for this purpose. It is a simple number scale based on the hydrogen ion concentration. Number 7 on the pH scale is the neutral point. Below 7, acidity increases, and the lower the number the greater the acidity. Above 7, alkalinity increases, and the greater the number the greater the alkalinity.

Feedwater should be alkaline rather than acid to prevent corrosion, but too high a degree of alkalinity may induce caustic embrittlement. Correct pH value can be determined only by careful tests in each case.

A simple test for finding pH value can be made by adding a definite amount of a prepared dye, called an *indicator,* to a measured sample of water, allowing it to stand for a few minutes, then matching the color of the mixture with a set of numbered standard color tubes. Dye indicators and color standards can be procured from any chemical supply house.

REMOVING IMPURITIES

Q What dangers may arise from oil passing into the boiler with feedwater?

A Oil will coat the heating surfaces and cause them to overheat, keeping the water from making direct contact with plates and tubes. Oil will also form a scum on the surface of the water and prevent steam bubbles from breaking away readily from the surface, thus causing foaming. Oil can be removed from the feedwater by proper filtration.

Q Why is it desirable that air and other dissolved gases be removed from feedwater before it enters the boiler?

A Some of these gases, particularly oxygen, have a very corrosive action on metal. Carbon dioxide may also produce corrosion or combine with other impurities to produce scale.

Q Describe an apparatus for removing air and other dissolved gases from boiler feedwater.

A If the water is heated close to the boiling point in an open heater, most of the dissolved gases can be separated out and vented to the atmosphere. For more complete oxygen removal, use a deaerator or deaerating heater.

An improved form of open heater which also acts as a very effective deaerator is shown in Fig. 15-8. Feedwater passes first through the tubes of a small vent condenser, then discharges into the upper part of the deaerator in the form of a great number of fine sprays. Entering steam mingles with these finely divided streams of water, heating the water and liberating the air and other dissolved gases. This process continues as the water trickles downward over banks of small trays, which still keep it in a finely divided state.

Most of the steam is condensed by contact with the water, but any remaining vapor passes with the separated air and gases to the vent condenser, where the steam is condensed by contact with the tubes containing the incoming water supply. The condensed steam drains back to the lower compartment of the deaerator, and the air and gases are vented to the atmosphere if the deaerator is being operated at a pressure above atmospheric pressure. If the pressure in the deaerator is below atmospheric pressure, a vacuum pump is used to draw off air and other gases from the vent condenser.

Q   What is the special purpose of an economizer in a steam-boiler plant?
A   An economizer is a divice for using some of the flue-gas heat that would otherwise go to waste by pumping boiler feedwater through banks of tubes placed in the direct path of gas passing to the chimney. Since the economizer is placed between feed pumps and boiler, water is under a

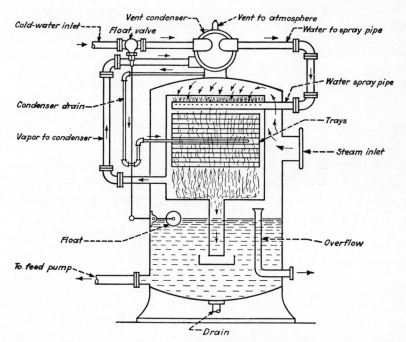

FIG. 15-8   Deaerating feed-water heater vents noncondensible gases to atmosphere.

little higher pressure than boiler pressure and can be heated to a much higher temperature than is possible when feedwater is heated at or near atmospheric pressure.

Cast-iron pipe was formerly used in practically all economizer construction because cast iron resists corrosion better than steel. This type of economizer is still suitable for low or medium pressures, but cast iron is not reliable under the higher boiler pressures now in common use. Therefore steel-pipe construction is standard practice in high-pressure boiler installations, with deaerators to rid feedwater of air and other corrosive gases before it enters the economizer. Tubes are kept free of soot on the outside by built-in soot blowers similar to those used in water-tube boiler settings.

Figure 15-9 shows a steel-tube economizer element fitted with a finned casting to increase the area of heat-absorbing surface. The tubes are expanded into steel cross-box connections, as shown, and there is a cleanout plug opposite each tube hole. For support only, the banks of tubes pass through holes in cast-iron plates. They are enclosed by cast-iron plates lined with insulating material.

Economizers offer considerable resistance to the passage of flue gas, and therefore some form of mechanical draft must be installed when they are used. They take up space, add to the initial cost of installation, and involve extra maintenance cost for cleaning and repair. On the other hand, economizers may effect quite a large saving in fuel consumption since a 50°F temperature increase saves about 5 percent of the fuel. They increase boiler efficiency substantially when boilers are operated at a high percentage over normal rating. Higher feed-water temperatures tend to reduce boiler stresses.

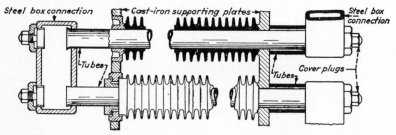

FIG. 15-9 Details of steel-tube economizer element that absorbs heat from exhaust gases flowing over outside.

SUGGESTED READING ────────────────────────────────

REPRINT SOLD BY *Power* MAGAZINE
Water Treatment, 24 pp.

BOOKS
Elonka, Stephen M., and Anthony L. Kohan: "Standard Boiler Operators' Questions and Answers," McGraw-Hill Book Company, New York, 1969.
Elonka, Stephen M.: "Standard Plant Operators' Manual," McGraw-Hill Book Company, New York, 1975.

16

HEAT, HOW TO USE STEAM TABLES, HOT WATER

Heat (energy) is the key to modern civilization. Yet our known sources of heat (fuel) are diminishing daily. Since it is the stationary engineer who is charged with operating equipment which releases this heat and turns it into energy, isn't it logical that the engineer should *know* as much about heat as possible to do the job efficiently?

Here we not only explain heat, but how any operating engineer can use the steam tables to figure out problems involving heating water, making steam, and superheating.

HEAT _____

Q What is heat?
A It is a form of energy due to molecular motion. The molecules of any substance containing heat are assumed to be constantly in motion, and the intensity (or temperature) of heat depends upon the rapidity of this molecular vibration. The temperature of a body will rise as the rate of vibration increases, and fall as it decreases. To understand this fully, just consider that at absolute zero ($-459.8°F$), there is *no* molecular action, therefore *no* heat. That means anything *above* absolute zero has its molecules in constant motion.

Q What are the effects of adding heat to a body?

A Addition of heat to a body may cause: (1) rise in temperature; (2) change of state—for example, from solid to liquid (ice to water) or liquid to gas (water to steam); (3) performance of external work by expansion of the solid, liquid, or gaseous body to which the heat is added.

All the foregoing effects are seen if heat is applied to ice to melt it into water, thus changing its state without raising its temperature; if addition of heat is continued until the water reaches the boiling point, thus raising its temperature without changing its state; finally, if heat is added until the water turns into steam, which is another change of state without a rise in temperature.

As steam is generated, it exerts pressure on the walls of the vessel in which it is confined. If this vessel is a cylinder containing a movable piston, the steam can do external work by moving the piston.

Q How is heat transferred from one body to another?

A It is convenient to think of heat as flowing like a fluid from one body to another, but strictly speaking there is no transfer of any physical substance. Molecules in the hotter substance are vibrating at a higher rate than those in the colder substance; therefore when the bodies are brought into contact, the effect is to increase molecular vibration in the colder body and decrease vibration in the hotter body until equilibrium is established. Unless artificially reversed by outside power (as in a refrigerating machine), heat transfer is always from the hotter to the colder body.

Q What is temperature?

A It is a measure of heat *intensity*, or *degree* of hotness or coldness as distinct from *quantity*. A very small body and a very large one may be at exactly the same temperature, but it is obvious that the large body contains a much greater *quantity* of heat than the small body.

MEASURING TEMPERATURE

Q What is meant by absolute temperature?

A The volume of a perfect gas, under constant pressure, is known to decrease $1/273$ of its volume at 0°C for every degree centigrade fall in temperature. From this it would appear that at 273° below zero on the centigrade scale, volume of the gas would be reduced to zero, and the molecular motion that produces heat would have entirely ceased. This extreme low point is called *absolute zero*, meaning the lowest possible temperature that could be attained. In the same way, 460° below zero on the Fahrenheit scale is the absolute zero for that scale.

Absolute temperatures are reckoned from absolute zero. To reduce any Fahrenheit reading to absolute temperature, add 460°, and to reduce any centigrade reading to absolute temperature, add 273°. Thus 26°F would be $26 + 460 = 486°F$ abs, and 26°C would be $26 + 273 = 299°C$ abs.

Q In what units is temperature measured, and what instruments are used to measure it?

A Temperature is measured in *degrees*. Thermometers measure all ordinary ranges of temperatures up to around 1000°F; pyrometers measure very high temperatures beyond the range of thermometers.

Q Describe the construction of a thermometer.

A A thermometer is a glass tube with a very small central bore, having one end blown into bulb form and the other end closed. Bulb and tube are partly filled with a liquid. (Mercury and alcohol are commonly used for this purpose.) The air is exhausted from the remaining portion of the tube, except in very high-temperature thermometers where this space is filled with a gas. Approximate ranges of the common types of glass thermometers are: mercury-filled, from −38 to 750°F; mercury- and nitrogen-filled, from −38 to 1000°F; alcohol-filled, from −95 to 150°F.

When placed in a heated atmosphere or liquid, mercury or alcohol expands and travels upward in the tube, a very small expansion causing quite a noticeable upward movement. A scale of degrees is etched on the glass of a mercury thermometer, marked as follows:

1. The thermometer is immersed in melting ice at a pressure of 14.7 psia and a mark is made at the top of the mercury column. This is the *freezing point*, called 0° on the centigrade scale and 32° on the Fahrenheit. The two thermometer scales are compared in Fig. 16-1.

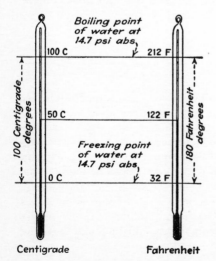

FIG. 16-1 Comparison of centigrade and Fahrenheit thermometer scales.

2. The thermometer is immersed in boiling water at a pressure of 14.7 psi abs and a mark made at the top of the mercury column. This is the *boiling point,* called 100° on the centigrade scale and 212° on the Fahrenheit.

3. Distance between the freezing and boiling points is divided into 100 equal parts, or 100°, on the centigrade thermometer, and into 180 equal parts, or 180°, on the Fahrenheit.

The centigrade thermometer has a more logical scale than the Fahrenheit, and is the one commonly used in scientific calculations, but the Fahrenheit is widely used by engineers and others for many everyday purposes.

Q How can thermometer readings in centigrade degrees be converted into readings in Fahrenheit degrees?
A Between the freezing and the boiling points there are 180° on the Fahrenheit scale and 100° on the centigrade scale; hence 180 Fahrenheit degrees equals 100 centigrade degrees. Dividing each by 10, we get a simpler ratio of 18 Fahrenheit degrees equals 10 centigrade degrees. So if we multiply centigrade degrees by 1.8, we get the corresponding number of Fahrenheit degrees. In order, however, that the readings may exactly correspond in position on both scales, we must add 32 because 32 on the Fahrenheit scale is equivalent to 0 on the centigrade scale. The rule is: to convert centigrade temperature to Fahrenheit, multiply by 1.8 and add 32. Thus:

$$22°C = 22 \times 1.8 + 32 = 39.6 + 32 = 71.6°F$$

Q How can you convert readings in Fahrenheit into centigrade degrees?
A To convert Fahrenheit into centigrade degrees, subtract 32, then divide by 1.8. Thus:

$$105°F = (105 - 32) \div 1.8 = 73 \div 1.8 = 40.5°C$$

Q What instruments are used for measuring very high temperatures?
A *Pyrometers* measure temperatures above the range of thermometers. There are a number of types, but those operated electrically are probably most common. Of these the *thermocouple* and *optical* pyrometers are typical.

In the thermocouple pyrometer, two rods or wires of dissimilar metals are joined and sealed in a porcelain tube, Fig. 16-2. Wires are connected to these rods and to a galvanometer. The tube containing the rods is exposed to the heat at the point where it is desired to measure the temperature. As the rods heat up, an electric voltage is induced at their junction, proportional to the difference in temperature between the hot junction and the so-called cold junction (where the rods connect to the leads). The resulting current flows through the circuit and deflects the galvanometer

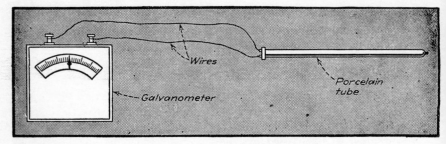

FIG. 16-2 Thermocouple pyrometer for measuring high temperatures.

needle. The dial of the galvanometer is graduated to read in degrees of temperature.

As for the optical pyrometer, it has a telescope containing a tiny filament that glows when an electric current passes through it. In the circuit with this filament is a small battery and a galvanometer. By means of a variable resistance built into the telescope, current flow through the filament is varied until the filament seems to disappear entirely when the telescope is focused on the furnace flame or wall. At this point the temperature is read on the galvanometer dial.

Unlike the thermocouple, no part of the optical pyrometer is exposed directly to furnace heat, and one can be quite a distance from the flame being observed. Operation of this instrument depends upon the fact that there is a definite relationship between color and temperature.

The electrical optical pyrometer has an electric battery, but none is needed with the thermocouple pyrometer.

Q How is quantity of heat measured?
A Quantity of heat is measured in British thermal units (usually abbreviated Btu).

Q What is a British thermal unit?
A One Btu is the $\frac{1}{180}$ part of the heat required to raise the temperature of one pound of water from 32°F to 212°F, or about the amount of heat needed to raise the temperature of one pound of water through one degree Fahrenheit. The latter definition is close enough for most practical purposes.

SPECIFIC HEAT

Q What is meant by the specific heat of a substance?
A In the English system of units, it is the amount of heat required to raise the temperature of one pound of that substance through one degree Fahrenheit.

Q Give the specific heats of some common materials.
A The following short table gives the specific heats of several very common substances:

Material	Specific heat Btu/lb/°F
Water	1.000
Ice	0.49
Iron (cast).	0.13
Copper	0.093
Aluminum	0.218
Brass.	0.088
Coal	0.318
Concrete	0.270
Air (at constant pressure).	0.24
Air (at constant volume)	0.17
Flue gas (at constant pressure).	0.240

Figures are approximate, as some materials vary greatly in composition. Also, specific heat changes with temperature.

Q How is heat transmitted from one substance to another?
A By radiation, conduction, and convection.

Q What is radiation?
A It is the giving off of heat from a hot body by ether waves of the same nature as light waves. Radiant heat does not warm the air to any great extent as it passes through it, but is absorbed or reflected by any solid obstruction. In a boiler furnace we have direct heat radiation from the boiler fire to all parts of the boiler that can "see" the fire.

Q What is conduction?
A It is the passage of heat through a body by the contact of one molecule with another. For example, if one end of an iron bar is placed in a fire while the other end is held in the hand, in a short time the end in the hand will become unbearably hot because of the conduction of heat through the bar from the red-hot end. Here heat is passed along by a series of collisions; fast-moving hot molecules bump into and speed up the cooler, slower molecules. Heat passes in this way through the tube walls and plates of a boiler to the water on the other side.

Q What is convection?
A Convection is the transfer of heat by current flow. As gases or liquids are heated by conduction through the walls of a containing vessel, they tend to expand and rise, and their place is taken by the upper colder layers of liquid or gas which, being heavier than the heated liquid or gas, tend to flow downward. In this way convection currents are set up, and

the whole body of gas or liquid is gradually heated to a uniform temperature.

It is by means of convection currents that the air of a room is heated to uniform temperature by a steam radiator. Water in a steam boiler is also heated uniformly throughout by convection currents set up by upward flow of the lighter heated water in contact with the heating surface, and by the downward flow of the heavier colder water above.

Q What is thermal conductivity?
A Thermal conductivity refers to the rate at which heat passes through a body. The rate varies widely for different substances, and may (but not necessarily) be stated as the number of Btu that can flow in one hour through a block of the material, one square foot in area and one inch thick, with 1°F difference in temperature between the opposite surfaces. Since thermal conductivity varies with temperature, density, and moisture content, tables of thermal conductivity give only very approximate values. The rate of conductivity of metals usually decreases as temperature rises, but for most other substances the rate increases as temperature rises.

Q Give the thermal conductivity of some common substances.
A Table 16-1 gives average values of thermal conductivity of some common materials at the temperatures noted.

Q What is the coefficient of linear expansion of a solid body?
A It is the fraction of the body's length that it expands when heated 1°F. Stated another way, it is the amount of expansion per unit length, per degree rise in temperature. For example, if the coefficient of steel is

TABLE 16-1 Thermal Conductivity or Heat Transfer
(Btu per sq ft, 1 in. thick, per °F temperature difference, per hr)

Substance	Temperature °F	Conductivity
Air	50	1.58
Water	140	4.62
Common brick	70	4.56
Firebrick	2000	12.00
Lead	64	241.00
Cast iron	216	320.00
Steel	212	310.00
Yellow brass	212	738.00
Copper	64	2668.00

TABLE 16-2   Coefficients of Linear Expansion
(Average values per °F between 32 and 212°F)

Aluminum	0.000,0128
Brass (cast)	0.000,0104
Bronze	0.000,0104
Copper	0.000,0091
Cast iron	0.000,0059
Wrought iron	0.000,0063
Lead	0.000,0164
Nickel	0.000,0072
Steel	0.000,0063
Tin	0.000,0119
Zinc	0.000,0219
Glass	0.000,0050

0.0000063, a piece of steel expands that fraction of its length for each degree rise in temperature. If it is 1 in. long, it expands 0.0000063 in.; and if it is 1 ft long, it expands 0.0000063 ft. To put it another way, steel expands 6.3 parts in a million for every degree rise.

Q Give the coefficients of linear expansion of some common materials.
A Coefficients of expansion of metals and alloys vary with composition and degree of purity. In some cases the coefficient increases at temperatures above 212°F, but the increase is small, and the values in Table 16-2 above give results close enough for all practical purposes.

Q What do we know about the expansion and contraction of liquids?
A Most liquids expand in volume when heated, and contract when cooled. They expand more than solids and exert enormous pressure when expanding if they are confined in a closed vessel. Different liquids have different rates of expansion. Ether, alcohol, and light oils such as gasoline have a much greater rate of expansion than water. For liquids, we measure *cubical* expansion, and the coefficient of cubical expansion is the expansion per unit volume, per degree Fahrenheit rise in temperature.

WATER AND GASES

Q In what way does the behavior of water under expansion and contraction differ from that of other liquids?
A If water is cooled it contracts in volume until a temperature of 39°F is reached. From 39°F down to 32°F it *expands*. At 32°F it freezes, the act of

freezing being accompanied by further expansion. Ice then is lighter than water, and water freezes first on the surface. If it were not for this fact, streams and lakes would freeze solid during severely cold weather, and all plant and animal life beneath the surface would be destroyed. It is the expansive force of water when freezing that bursts water pipes and tanks.

Q How do gases behave when heated?

A Gases increase in volume or pressure when heated, and decrease in volume or pressure when cooled, according to two simple laws, Charles's law and Gay-Lussac's law. When considering problems involving these two laws, it is also customary to study another simple law, Boyle's law, which deals with pressure and volume changes, as changes of temperature are often accompanied by changes of pressure.

Q What is Boyle's law?

A Boyle's law states that "the absolute pressure of a gas will vary *inversely* as the volume, if the temperature remains constant." Or conversely, "The volume will vary *inversely* as the absolute pressure, if the temperature remains constant."

According to this, if pressure is increased, volume is decreased proportionally, or vice versa. For example, if we have 10 cu ft of gas under a pressure of 10 psia and we increase the pressure to 20 psia, the volume will be reduced to 5 cu ft. Doubling the one halves the other for constant temperature.

Q What is Charles's law?

A This law states that "the absolute pressure of a gas will vary *directly* as the absolute temperature, if the volume remains constant." Notice that this is a *direct* proportion. If the absolute temperature rises 30 percent, the absolute pressure rises 30 percent.

Q What is the law of Gay-Lussac?

A The law states, "The volume of a gas will vary *directly* as the absolute temperature, if the pressure remains constant."

Q What is meant by *isothermal* expansion or compression?

A It is the expansion or compression of a gas at constant temperature, that is, with the temperature remaining the same during expansion or compression. This is the condition that would exist if the change were taking place according to Boyle's law.

In practice, we never have true isothermal expansion or compression. Even a water-jacketed compressor cylinder can't carry heat away fast enough, and therefore air temperature rises sharply during compression.

Q What is *adiabatic* expansion or compression?

A This is expansion or compression where the temperature rises during compression and falls during expansion without any loss of heat to the cylinder walls or absorption of heat from the walls. This condition is never exactly realized in practice, although it is approached fairly closely in some gas engines and air compressors.

Q If the original volume of a gas is 40 cu ft at a temperature of 60°F, what volume will it occupy if the temperature is increased to 190°F, the pressure remaining constant?
A According to Gay-Lussac's law, volume varies directly as the absolute temperature if the pressure remains constant. Converting to absolute temperatures and applying this law:

$$60 + 460 = 520°F \text{ abs}$$
$$190 + 460 = 650°F \text{ abs}$$
$$\text{Temperature ratio} = 650 \div 520 = 1.25$$
$$\text{Final volume} = 1.25 \times 40 = 50 \text{ cu ft} \quad Ans.$$

Q If 4 cu ft of air at atmospheric pressure is to be compressed to 100 psi gage, at constant temperature, what volume will the air then occupy?
A This is the condition assumed in Boyle's law, which tells us that volume varies inversely as the absolute pressure, temperature remaining constant. Converting to absolute pressures and applying Boyle's law:

$$14.7 + 0 = 14.7 \text{ psia}$$
$$100.0 + 14.7 = 114.7 \text{ psia}$$
$$\text{Volume ratio} = 14.7 \div 114.7 = 0.128$$
$$\text{Final volume} = 4 \times 0.128 = 0.512 \text{ cu ft} \quad Ans.$$

Q A given weight of gas is at atmospheric pressure and a temperature of 60°F. If its volume is 40 cu ft, what volume will it occupy if the temperature is raised to 180°F and the pressure is raised to 80 psi?
A As this problem involves changes of temperature, pressure, and volume, we must use more than one of the simple gas laws in its solution.
Considering change of volume due to pressure change only, and applying Boyle's law, we find:

$$14.7 + 0 = 14.7 \text{ psia}$$
$$80 + 14.7 = 94.7 \text{ psia}$$
$$\text{Volume ratio} = 14.7 \div 94.7 = 0.155$$
$$\text{Final volume} = 40 \times 0.155 = 6.2 \text{ cu ft} \quad Ans.$$

Considering change of volume due to temperature change only, and applying Gay-Lussac's law, we find:

$$60 + 460 = 520°F \text{ abs}$$
$$180 + 460 = 640°F \text{ abs}$$
$$\text{Temperature ratio} = 640 \div 520 = 1.23$$
$$\text{Final volume} = 6.2 \times 1.23 = 7.63 \text{ cu ft} \quad Ans.$$

STEAM TABLES

Q What is steam?

A Steam is water in a semigaseous condition. Though it obeys to some extent the simple laws governing the behavior of true gases, it is a *vapor* rather than a gas—that is, a substance between the purely liquid and gaseous forms.

Q Explain how steam is generated from water in a steam boiler.

A Furnace heat is conducted through the metal of plates and tubes, heating the water directly in contact with the metal. This heated water, being now lighter than the colder water above, rises, and the colder water flows downward to take its place. By the convection currents thus set up, the whole mass of water is gradually heated to the boiling point; further addition of heat changes it into steam. There is no change of weight in the process; one pound of water changes into one pound of steam.

Q Upon what does the boiling point of water depend?

A It depends upon the pressure exerted on the water. At sea level the boiling point of water under atmospheric pressure (14.7 psia) is 212°F. As pressure decreases, the boiling point decreases, and, as pressure increases, the boiling point increases. Thus at 10 psia, the boiling point is 193.21°F. At 20 psia, the boiling point is 227.96°F.

Q In connection with the generation of steam, what is meant by *sensible* heat?

A It is heat added to water to bring it from 32°F to the boiling point. Temperature rise can be measured on a thermometer and felt by the physical senses. Hence the name, *sensible* heat.

Q What is the *latent* heat of steam?

A This is the amount of heat required to convert water at the boiling point into steam at the same temperature and pressure. The word *latent* means hidden, and is applied here because there is no indication of the effect of adding heat other than the change in state from a liquid to a vapor.

Q What is the *total* heat of steam?

A This is the sum of the sensible heat and latent heat. The latest steam tables use the word *enthalpy* to replace heat in the expressions just defined. Thus sensible heat becomes *enthalpy of the liquid,* latent heat becomes *enthalpy of evaporation,* and total heat, *enthalpy of steam.* Water or steam at the boiling point is said to be saturated. Table 16-3 uses the abbreviations *sat. liquid* for heat of liquid or sensible heat, *evap.* for latent heat, and *sat. vapor* for total heat, all under the general heading *enthalpy.*

TABLE 16-3 Properties of Saturated Steam

Abs. press., psi	Temp., °F	Specific vol. Sat. liquid	Specific vol. Sat. vapor	Enthalpy Sat. liquid	Enthalpy Evap.	Enthalpy Sat. vapor
0.50	79.58	0.01608	641.4	47.6	1048.8	1096.4
1.0	101.74	0.01614	333.6	69.7	1036.3	1106.0
2.0	126.08	0.01623	173.73	94.0	1022.2	1116.2
3.0	141.48	0.01630	118.71	109.4	1013.2	1122.6
4.0	152.97	0.01636	90.63	120.9	1006.4	1127.3
5.0	162.24	0.01640	73.52	130.1	1001.0	1131.1
6.0	170.06	0.01645	61.98	138.0	996.2	1134.2
7.0	176.85	0.01649	53.64	144.8	992.1	1136.9
8.0	182.86	0.01653	47.34	150.8	988.5	1139.3
9.0	188.28	0.01656	42.40	156.2	985.2	1141.4
10	193.21	0.01659	38.42	161.2	982.1	1143.3
14.7	212.00	0.01672	26.80	180.0	970.3	1150.4
15	213.03	0.01672	26.29	181.1	969.7	1150.8
20	227.96	0.01683	20.089	196.2	960.1	1156.3
25	240.07	0.01692	16.303	208.4	952.1	1160.6
30	250.33	0.01701	13.746	218.8	945.3	1164.1
40	267.25	0.01715	10.498	236.0	933.7	1169.7
50	281.01	0.01727	8.515	250.1	924.0	1174.1
60	292.71	0.01738	7.175	262.1	915.5	1177.6
70	302.92	0.01748	6.206	272.6	907.9	1180.6
80	312.03	0.01757	5.472	282.0	901.1	1183.1
90	320.27	0.01766	4.896	290.6	894.7	1185.3
100	327.81	0.01774	4.432	298.4	888.8	1187.2
110	334.77	0.01782	4.049	305.7	883.2	1188.9
120	341.25	0.01789	3.728	312.4	877.9	1190.4
130	347.32	0.01796	3.455	318.8	872.9	1191.7
140	353.02	0.01802	3.220	324.8	868.2	1193.0
150	358.42	0.01809	3.015	330.5	863.6	1194.1
160	363.53	0.01815	2.834	335.9	859.2	1195.1
170	368.41	0.01822	2.675	341.1	854.9	1196.0
180	373.06	0.01827	2.532	346.0	850.8	1196.3
190	377.51	0.01833	2.404	350.8	846.8	1197.6
200	381.79	0.01839	2.288	355.4	843.0	1198.4
250	400.95	0.01865	1.8438	376.0	825.1	1201.1
300	417.33	0.01890	1.5433	393.8	809.0	1202.8
350	431.72	0.01913	1.3260	409.7	794.2	1203.9
400	444.59	0.0193	1.1613	424.0	780.5	1204.5
450	456.28	0.0195	1.0320	437.2	767.4	1204.6
500	467.01	0.0197	0.9278	449.4	755.0	1204.4
600	486.21	0.0201	0.7698	471.6	731.6	1203.2
700	503.10	0.0205	0.6554	491.5	709.7	1201.2
800	518.23	0.0209	0.5687	509.7	688.9	1198.6
900	531.98	0.0212	0.5006	526.6	668.8	1195.4
1000	544.61	0.0216	0.4456	542.4	649.4	1191.8
1100	556.31	0.0220	0.4001	557.4	630.4	1187.7
1200	567.22	0.0223	0.3619	571.7	611.7	1183.4
1300	577.46	0.0227	0.3293	585.4	593.2	1178.6
1400	587.10	0.0231	0.3012	598.7	574.7	1173.4
1500	596.23	0.0235	0.2765	611.6	556.3	1167.9
2000	635.82	0.0257	0.1878	671.7	463.4	1135.1
2500	668.13	0.0287	0.1307	730.6	360.5	1091.0
3000	695.36	0.0346	0.0858	802.5	217.8	1020.3
3206.2	705.40	0.0503	0.0503	902.7	0	902.7

Condensation, by permission, from Keenan and Keyes, "Thermodynamic Properties of Steam," John Wiley & Sons, Inc.
NOTE: The ASME (American Society of Mechanical Engineers) also publishes steam tables.

Q What is *saturated* steam?

A It is steam as it is generated from water, just barely on the steam side of the fence, so to speak. Any loss of heat without a corresponding drop in pressure immediately starts to condense it back into water.

Q What is *dry saturated* steam?

A If the saturated steam, as generated from water, contains no moisture in suspension (that is, no fine drops of liquid water as in fog) it is said to be dry. If it contains moisture it is called wet steam. Completely dry steam is invisible. The white, foggy appearance of most steam discharged to atmosphere is caused by particles of liquid water in suspension.

Q What is meant by *quality* of steam?

A The term *quality* refers indirectly to the amount of water or unevaporated moisture in steam. If it is perfectly dry, its quality will be 100 percent, but if it contains, say, 2 percent of moisture, its quality will be $100 - 2 = 98$ percent.

Q How is the quality of steam determined?

A It is determined by means of an apparatus called a *calorimeter*. There are three types: the *barrel* calorimeter, a primitive type and not very accurate; the *throttling* calorimeter, which determines percentages up to about 7 percent at 400 psi gage; and the *separating* calorimeter, which has a wider range and is as a rule more accurate than the first two.

Q What is *superheated* steam, and why is steam superheated?

A It is steam at a temperature higher than the saturation temperature for the given pressure. Steam is superheated by passing saturated steam through tube coils exposed to furnace heat. Steam thus raised in temperature, or superheated, will have to drop in temperature by the amount of its superheat before it begins to condense. This is of considerable benefit in power-plant operation, as it allows for radiation loss of heat in steam piping, lessens danger of damage from water hammer in pipe lines and engine cylinders, improves thermal efficiency of engines and turbines, and reduces the bad effects of excessive moisture in the low-pressure stages of steam turbines.

Q What are saturated-steam tables?

A They are tabulated values of various properties of saturated steam such as boiling point (saturation temperature), specific volume (volume of 1 lb in cu ft), sensible heat (enthalpy of saturated liquid), latent heat (enthalpy of evaporation), and total heat (enthalpy of saturated vapor or of superheated vapor), calculated for a wide range of pressures. Absolute pressures are used in steam tables because gage pressures are based on atmospheric pressure, which varies with altitude and weather conditions.

All quantities in Tables 16-3 and 16-4 refer to 1 lb of water or dry and saturated steam. Temperatures are in degrees Fahrenheit. Enthalpy is in Btu. Enthalpy of the liquid or of steam is the amount of heat required to produce the given water or steam, starting with water at 32°F.

At atmospheric pressure (14.7 psia) enthalpy of the saturated liquid is $212 - 32 = 180$ Btu, or temperature of the water less 32.

Q What does a study of steam tables show?

A The tables show that each pressure has a corresponding boiling point and, as the pressure rises, the following changes take place: (1) boiling-point temperature rises, (2) sensible heat increases, (3) latent heat decreases, (4) total heat increases slowly until pressure is approximately 450 psia, then decreases slowly until pressure approaches the so-called *critical point,* of 3,206.2 psia.

Q What would be the effect of a sudden rise or a sudden drop in pressure in a closed vessel containing steam if there were no temperature change?

A Pressure rise without a corresponding temperature rise would cause some of the steam to condense, as the temperature would then be below the boiling point corresponding to the pressure. A drop in pressure would cause the steam to become superheated because its temperature would then be above the boiling-point temperature corresponding to the pressure.

If the pressure drop takes place in a boiler where the steam is in contact with the water from which it is generated, some of the water will evaporate into steam because of the lowered boiling point. If the lowering of pressure is brought about very suddenly by the rupture of some part of the boiler, much of the water in the boiler may instantly flash into steam, causing a disastrous explosion. Since 1 lb of steam at atmospheric pressure occupies about 1,600 times the space occupied by 1 lb of water, the explosive energy released by such a sudden pressure drop would be tremendous.

Q Compare the use of steam for power and for heating purposes.

A An examination of Tables 16-3 and 16-4 shows that the greater part of the total heat necessary to convert water into steam is used up, not in raising temperature but in changing the boiling liquid into a vapor, and this latent heat is not given back by the steam unless it is condensed into water.

In a *heating system,* the steam condenses into water in the radiators or heating coils, and, in doing so, gives up its latent heat. Thus the greater part of the heat in steam does useful work when steam is used for heating purposes, and the system therefore has a high efficiency.

TABLE 16-4 Properties of Superheated Steam

Abs. press, psi (sat. temp.)*		Sat. liquid	Sat. vapor	Temperature, °F							
				300	400	500	600	700	800	900	1000
15 (213.03)	v	0.02	26.29	29.91	33.97	37.99	41.99	45.98	49.97	53.95	57.93
	h	181.1	1150.8	1192.8	1239.9	1287.1	1334.8	1383.1	1432.3	1482.3	1533.1
	s	0.3135	1.7549	1.8136	1.8719	1.9238	1.9711	2.0147	2.0554	2.0936	2.1296
20 (227.96)	v	0.02	20.09	22.36	25.43	28.46	31.47	34.47	37.46	40.45	43.44
	h	196.2	1156.3	1191.6	1239.2	1286.6	1334.4	1382.9	1432.1	1482.1	1533.0
	s	0.3356	1.7319	1.7808	1.8396	1.8918	1.9392	1.9829	2.0235	2.0618	2.0978
40 (267.25)	v	0.017	10.498	11.040	12.628	14.168	15.688	17.198	18.702	20.20	21.70
	h	236.0	1169.7	1186.8	1236.5	1284.8	1333.1	1381.9	1431.3	1481.4	1532.4
	s	0.3919	1.6763	1.6994	1.7608	1.8140	1.8619	1.9058	1.9467	1.9850	2.0212
60 (292.71)	v	0.017	7.175	7.259	8.357	9.403	10.427	11.441	12.449	13.452	14.454
	h	262.1	1177.6	1181.6	1233.6	1283.0	1331.8	1380.9	1430.5	1480.8	1531.9
	s	0.4270	1.6438	1.6492	1.7135	1.7678	1.8162	1.8605	1.9015	1.9400	1.9762
80 (312.03)	v	0.018	5.472		6.220	7.020	7.797	8.562	9.322	10.077	10.830
	h	282.0	1183.1		1230.7	1281.1	1330.5	1379.9	1429.7	1480.1	1531.3
	s	0.4531	1.6207		1.6791	1.7346	1.7836	1.8281	1.8694	1.9079	1.9442
100 (327.81)	v	0.018	4.432		4.937	5.589	6.218	6.835	7.446	8.052	8.656
	h	298.4	1187.2		1227.6	1279.1	1329.1	1378.9	1428.9	1479.5	1530.8
	s	0.4740	1.6026		1.6518	1.7085	1.7581	1.8029	1.8443	1.8829	1.9193
120 (341.25)	v	0.018	3.728		4.081	4.636	5.165	5.683	6.195	6.702	7.207
	h	312.4	1190.4		1224.4	1277.2	1327.7	1377.8	1428.1	1478.8	1530.2
	s	0.4916	1.5878		1.6287	1.6869	1.7370	1.7822	1.8237	1.8625	1.8990
140 (353.02)	v	0.018	3.220		3.468	3.954	4.413	4.861	5.301	5.738	6.172
	h	324.8	1193.0		1221.1	1275.2	1326.4	1376.8	1427.3	1478.2	1529.7
	s	0.5069	1.5751		1.6087	1.6683	1.7190	1.7645	1.8063	1.8451	1.8817

TABLE 16-4 *(Continued)*

Press. psia (t sat)											
160 (363.53)	v	0.018	2.834		3.008	3.443	3.849	4.244	4.631	5.015	5.396
	h	335.9	1195.1		1217.6	1273.1	1325.0	1375.7	1426.4	1477.5	1529.1
	s	0.5204	1.5650		1.5908	1.6519	1.7033	1.7491	1.7911	1.8301	1.8667
180 (373.06)	v	0.018	2.532		2.649	3.044	3.411	3.764	4.110	4.452	4.792
	h	346.0	1196.9		1214.0	1271.0	1323.5	1374.7	1425.6	1476.8	1528.6
	s	0.5325	1.5542		1.5745	1.6373	1.6894	1.7355	1.7776	1.8167	1.8534
200 (381.79)	v	0.018	2.288		2.361	2.726	3.060	3.380	3.693	4.002	4.309
	h	355.4	1198.4		1210.3	1268.9	1322.1	1373.6	1424.8	1476.2	1528.0
	s	0.5435	1.5453		1.5594	1.6240	1.6767	1.7232	1.7655	1.8048	1.8415
250 (400.95)	v	0.0187	1.8438			2.151	2.427	2.688	2.942	3.192	3.439
	h	376.0	1201.1			1263.4	1318.5	1371.0	1422.7	1474.5	1526.6
	s	0.5675	1.5263			1.5949	1.6495	1.6969	1.7379	1.7793	1.8162
300 (417.33)	v	0.0189	1.5433			1.7675	2.005	2.227	2.442	2.652	2.859
	h	393.8	1202.8			1257.6	1314.7	1368.3	1420.6	1472.8	1525.2
	s	0.5879	1.5104			1.5701	1.6268	1.6751	1.7184	1.7582	1.7954
400 (444.59)	v	0.0193	1.1613			1.2851	1.4770	1.6508	1.8161	1.9767	2.134
	h	424.0	1204.5			1245.1	1306.9	1362.7	1416.4	1469.4	1522.4
	s	0.6214	1.4844			1.5281	1.5894	1.6398	1.6842	1.7247	1.7623
500 (467.01)	v	0.0197	0.9278			0.9927	1.1591	1.3044	1.4405	1.5715	1.6996
	h	449.4	1204.4			1231.3	1298.6	1357.0	1412.1	1466.0	1519.6
	s	0.6487	1.4634			1.4919	1.5588	1.6115	1.6571	1.6982	1.7363
600 (486.21)	v	0.0201	0.7698			0.7947	0.9463	1.0732	1.1899	1.3013	1.4096
	h	471.6	1203.2			1215.7	1289.9	1351.1	1407.7	1462.5	1516.7
	s	0.6720	1.4454			1.4586	1.5323	1.5875	1.6343	1.6762	1.7147
800 (518.23)	v	0.0209	0.5687				0.6779	0.7833	0.8763	0.9633	1.0470
	h	509.7	1198.6				1270.7	1338.6	1398.6	1455.4	1511.0
	s	0.7108	1.4153				1.4863	1.5476	1.5972	1.6407	1.6801

TABLE 16-4 (Continued)

Abs. press, psi (sat. temp.)*		Sat. liquid	Sat. vapor	Temperature, °F							
				300	400	500	600	700	800	900	1000
1,000 (544.61)	v	0.0216	0.4456				0.5140	0.6084	0.6878	0.7604	0.8294
	h	542.4	1191.8				1248.8	1325.3	1389.2	1448.2	1505.1
	s	0.7430	1.3897				1.4450	1.5141	1.5670	1.6121	1.6525
1,200 (567.22)	v	0.0223	0.3619				0.4016	0.4909	0.5617	0.6250	0.6843
	h	571.7	1183.4				1223.5	1311.0	1379.3	1440.7	1499.2
	s	0.7711	1.3667				1.4052	1.4843	1.5409	1.5879	1.6293
1,400 (587.10)	v	0.0231	0.3012				0.3174	0.4062	0.4714	0.5281	0.5805
	h	598.7	1173.4				1193.0	1295.5	1369.1	1433.1	1493.2
	s	0.7963	1.3454				1.3639	1.4567	1.5177	1.5666	1.6093
1,500 (596.23)	v	0.0235	0.2765				0.2815	0.3719	0.4352	0.4893	0.5390
	h	611.6	1167.9				1174.5	1287.2	1363.8	1429.3	1490.1
	s	0.8082	1.3351				1.3412	1.4434	1.5068	1.5569	1.6001
2,000 (635.82)	v	0.0257	0.1878					0.2489	0.3074	0.3532	0.3935
	h	671.7	1135.1					1240.0	1335.5	1409.2	1474.5
	s	0.8619	1.2849					1.3783	1.4576	1.5139	1.5603
2,500 (668.13)	v	0.0287	0.1307					0.1686	0.2294	0.2710	0.3061
	h	730.6	1091.1					1176.8	1303.6	1387.8	1458.4
	s	0.9126	1.2322					1.3073	1.4127	1.4772	1.5273
3,000 (695.36)	v	0.0346	0.0858					0.0984	0.1760	0.2159	0.2476
	h	802.5	1020.3					1060.7	1267.2	1365.0	1441.8
	s	0.9731	1.1615					1.1966	1.3690	1.4439	1.4984
3,206.2 (705.40)	v	0.0503	0.0503						0.1583	0.1981	0.2288
	h	902.7	902.7						1250.5	1355.2	1434.7
	s	1.0580	1.0580						1.3508	1.4309	1.4874

296

In a *power plant,* where steam is used for power purposes, the turbine exhaust is still in the form of steam, and contains the greater part of the original quantity of heat in the steam.

Q What are the critical temperature and pressure of steam?
A If steam is generated under conditions where pressures do not exceed several hundred psi, and the pressure could be suddenly raised a few pounds without a corresponding rise in temperature, some of the steam would be condensed or liquefied because the boiling temperature would now be higher. If, however, the pressure were allowed to rise to 3,206.2 psia, with a corresponding temperature of 705.4°F, no additional amount of pressure would cause the steam to condense. At this point the density of water and steam is the *same,* and latent heat of evaporation disappears entirely. Hence, if water under a pressure of 3,206.2 psia is heated to 705.4°F, it changes into steam there without further application of heat. These two values, 3,206.2 psia and 705.4°F, are known as the *critical pressure* and *critical temperature* of steam. It is impossible for water to exist as a liquid at any temperature above 705.4°F.

Q What is equivalent evaporation?
A All boilers do not work under the same conditions of pressure and feed-water temperature, so actual weight of water evaporated per hour or per pound of coal fired is not a fair way of comparing various boilers, unless pressure and temperature conditions are exactly alike. However, we can form a basis of comparison for any conditions of pressure and feed-water temperature by reducing each boiler's performance to a standard condition. One such condition is *the amount of water in pounds that would be evaporated from water at 212°F into steam at 212°F and 14.7 psia, by the heat put into the steam actually evaporated in 1 hr by 1 lb of fuel.* This amount is known as *equivalent evaporation,* from and at 212°F per lb of fuel.

The amount of heat to evaporate 1 lb of water from and at 212°F is 970.3 Btu. If we divide total heat put into the steam generated per hour, or per pound of fuel, by 970.3, we get the equivalent evaporation from and at 212°F per hr, or per lb, as the case may be.

Q A boiler takes feedwater at 175°F. Working steam pressure is 140 psia. Water evaporated per lb fuel burned is 6.8 lb. What is equivalent evaporation?
A Let H = total heat to evaporate 1 lb steam under these conditions,

$h = 1193.0$ Btu (total heat in 1 lb steam at 140 psia)
$t = 175°$F (from question)

Then

$$H = h - (t - 32) = 1193.0 - (175 - 32) = 1050.0 \text{ Btu}$$

Total heat per lb fuel burned $= 1050.0 \times 6.8 = 7140$ Btu

Equivalent evaporation $= 7140 \div 970.3 = 7.36$ lb water per lb fuel burned   *Ans.*

Q What is factor of evaporation?

A The equivalent evaporation that could be secured by the heat required to evaporate 1 lb steam at any given set of conditions is called the factor of evaporation for those particular conditions. Thus, in the previous question, the heat to evaporate 1 lb of water into steam was 1050.0 Btu (from feedwater at 175°F into steam at 140 psia). In that case,

$$\text{Factor of evaporation} = \frac{1050.0}{970.3} = 1.082 \quad Ans.$$

Factors of evaporation for various feed-water and steam-pressure conditions are given in Table 16-5. Other factors can readily be calculated from Tables 16-3 and 16-4.

Q What is a desuperheater, and when it is used?

A A desuperheater is an apparatus that brings superheated steam back to the saturated state by adding sufficient moisture to reduce its temperature to that corresponding to the temperature of dry and saturated steam at that particular pressure. It is used where all boilers are supplying superheated steam to a main header and it is desired to secure a supply of saturated steam from the header for some purpose.

Figure 16-3 shows the principle of one type of desuperheater. It is inserted in the superheated steam line as shown, and water is sprayed in the path of the flowing superheated steam to change it to the saturated form. A thermostat is placed in the line on the saturated-steam side of the desuperheater, and it controls the water supply to the water sprays in

TABLE 16-5   **Factors of Evaporation**

Steam pressure, psia	Temperature of feed water, °F						
	80	100	120	140	160	180	200
100	1.173	1.153	1.132	1.112	1.091	1.070	1.050
120	1.177	1.156	1.136	1.115	1.094	1.074	1.053
140	1.180	1.159	1.139	1.118	1.098	1.077	1.056
160	1.182	1.162	1.141	1.120	1.100	1.079	1.059
180	1.184	1.163	1.143	1.122	1.102	1.081	1.060
200	1.186	1.165	1.144	1.124	1.103	1.082	1.062

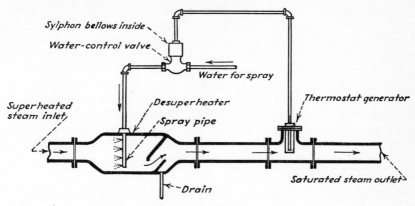

Sylphon bellows inside

Water-control valve

Water for spray

Superheated steam inlet

Desuperheater

Spray pipe

Thermostat generator

Drain

Saturated steam outlet

FIG. 16-3 Desuperheater for converting superheated steam into saturated steam.

the desuperheater. When the saturated-steam temperature rises above that corresponding to its pressure as dry and saturated steam, part of the volatile liquid in the generator of the thermostat is evaporated, thus creating a pressure in the line connecting it to the sylphon bellows on the water-control valve. This pressure expands the bellows, causing it to open the water-supply valve wider. When the temperature of the saturated steam falls below normal, liquid in the generator condenses, pressure in the connecting pipe line and sylphon bellows falls, and the bellows contracts and partly closes the water valve.

Q How are steam boilers rated?

A Sometimes some nominal horsepower rating based on square feet of heating surface is adopted for license, certificate, or other purposes; but because of differences in design and operating conditions, boilers with the same heating surface area may vary widely in evaporating capacity, which is the true measure of boiler power.

One method of rating boilers, based on actual performance, is the amount of water evaporated per square foot of heating surface per hour, reduced to equivalent evaporation. Another method takes heat output in steam, expressed in Btu per hour. A third method, based on evaporation, takes *1 boiler hp (actual output) as the evaporation of 34.5 lb of water per hr, from water at a temperature of 212°F, into steam at 212°F and 14.7 psia.* To find the actual horsepower output by this last method, reduce actual evaporation in pounds of water per hour to equivalent evaporation and divide by 34.5. This boiler horsepower has no connection with engine horsepower (mechanical horsepower). This method is still used as the common measure of capacity for small boilers.

Larger boiler capacity is almost invariably given in pounds of steam evaporated per hour, with the steam conditions specified. Maximum continuous rating is the hourly evaporation that can be maintained for 24 hr, and a further rating on the basis of a 2-hr peak output may be given.

A recent trend is toward rating large public utility boilers in kW (kilowatts), or mW (megawatts) of the turbine generator, thus including work done by the reheater. (In reheat turbines, steam expands in the front end of the turbine. At an intermediate state, the entire flow passes through a resuperheater in the boiler furnace. The reheated steam then goes back to the turbine at the lower pressure and expands through the turbine's remaining stages on its way to the condenser.)

Q Why is a steam generator not 100 percent efficient?
A It would be if it absorbed the entire heating value of the fuel burned. Unfortunately, the fuel-burning equipment and furnace cannot develop the full value of the fuel, and thus the steam generator proper starts with one strike against it. Theoretically, the steam generator efficiency is the ratio of the heat absorbed by the feedwater to the heat released in the furnace. Because it is difficult in practice to separate furnace losses from steam-generator losses and because, in the last analysis, it is overall efficiency that is of interest, it is customary to express steam-generating unit efficiency as the ratio of heat output in the steam to heat input in the fuel.

Q What is the developed horsepower output of a steam boiler working at 160 psia steam pressure and evaporating 3,400 lb of water per hr from feedwater at 140°F?
A Heat put into 1 lb of steam at 160 psia, with feedwater at 140°F, is $1195.1 - (140 - 32) = 1087.1$ Btu. Then

$$\text{Equivalent evaporation} = \frac{3,400 \times 1,087.1}{970.3}$$

$$= 3,809 \text{ lb of water per hr}$$
$$\text{Boiler hp} = 3,809 \div 34.5 = 110 \text{ hp} \quad Ans.$$

You shorten the work by using a table of factors of evaporation; thus the factor of evaporation here, from Table 16-5, is 1.12. Then

$$\text{Equivalent evaporation} = 3,400 \times 1.12 = 3,808 \text{ lb of water per hr}$$
$$\text{Boiler hp} = 3,808 \div 34.5 = 110 \text{ hp} \quad Ans.$$

Q What information is required for a simple boiler efficiency test?
A The following data must be secured: (1) temperature of feedwater, °F; (2) steam pressure, psi; (3) heat value of fuel used, Btu per lb; (4) water evaporated during test, lb; (5) fuel burned during test, lb; (6) duration of test, hr.

Q Find the efficiency of a steam boiler, given the following data:
Average feed-water temperature, 160°F
Steam pressure, 125 psig (approx. 140 psia)
Heat value of fuel, 10,200 Btu per lb
Water evaporated in test, 36,000 lb
Coal fired, 6,000 lb.
Duration of test, 8 hr

A Water evaporated per lb coal = 36,000 ÷ 6,000 = 6 lb.

Total heat put into 1 lb of steam = $h - (t - 32)$
$$= 1193 - (160 - 32) = 1065 \text{ Btu}$$
Total heat put into steam produced per lb of coal fired = 6 × 1065
$$= 6390 \text{ Btu}$$

$$\text{Efficiency of boiler} = \frac{\text{total heat in steam in Btu}}{\text{heat value of 1 lb fuel in Btu}}$$
$$= 6,390 \div 10,200 = 62.6 \text{ percent} \quad Ans.$$

Q If for licensing or other purposes, 10 sq ft of heating surface is taken as equivalent to 1 boiler hp, what would be the rating on this basis of an hrt boiler having the following dimensions: diameter 60 in., length 16 ft, outside tube diameter 3½ in., thickness of tube wall 0.12 in., number of tubes 60? Take lower half of shell, inner surface (gas-contact surface) of tubes, and two-thirds of area of the tube sheets, less area of tube holes, as heating surface.

A Heating surface of shell = $\dfrac{5 \times 3.142 \times 16}{2} = 125.7$ sq ft

Inside diameter of tubes = $3.5 - (2 \times 0.12) = 3.26$ in.

Area of tubes = $\dfrac{3.26 \times 3.142 \times 16 \times 60}{12} = 819.4$ sq ft

Area of tube sheets
$$= \frac{2}{3} \left(5 \times 5 \times 0.785 - \frac{60 \times 3.26 \times 3.26 \times 0.785}{144} \right) \times 2$$
$$= \frac{2}{3} (19.6 - 3.5) \times 2 = 21.5 \text{ sq ft}$$

Total heating surface = 125.7 + 8.19.4 + 21.5 = 966.6 sq ft
Nominal boiler hp = 966.6 ÷ 10 = 97 hp   *Ans.*

Q What is the specific heat of water?

A It is the heat in Btu required to raise the temperature of the water 1°F. The mean specific heat at atmospheric pressure is ¹/₁₈₀ of the heat required to raise the temperature of 1 lb of water through the 180° from 32 to 212°F. It is sufficiently accurate to consider this as 1. Even at higher temperatures, this is usually close enough for practical purposes.

Q What is the specific heat of superheated steam?
A The specific heat of superheated steam is the amount of heat required to raise the temperature of 1 lb of superheated steam, at constant pressure, 1°F. This amount varies with pressure and temperature, and the *mean specific heat* is found for any given set of conditions by dividing the increase in heat in Btu, as found from the superheated steam tables, by the increase in temperature in degrees Fahrenheit. For example, assume we wish to find the mean specific heat of superheat for steam at 200 psia at a temperature of 500°F. From the superheated steam tables, the temperature of saturated steam at 200 psia is 381.79°F, and the total heat is 1198.4 Btu. The total heat of the superheated steam is 1,268.9 Btu. Then

$$\text{Rise in temperature} = 500 - 381.79 = 118.21°F$$
$$\text{Increase in total heat} = 1268.9 - 1198.4 = 70.5 \text{ Btu}$$
$$\text{Mean specific heat} = \frac{70.5}{118.21} = 0.597 \quad Ans.$$

Q How much heat is required to produce 1 lb of superheated steam at 300 psia and 600°F from feedwater at 180°F?
A From the superheat table Table 16-4, final heat content (enthalpy) = 1314.7 Btu. Then

Heat added = 1314.7 − (180 − 32)
           = 1314.7 − 148
           = 1166.7 Btu per lb   *Ans.*

THE CALORIMETER

Q Describe a throttling calorimeter.
A Figure 16-4 shows construction of a common form. It has a cylindrical shell with a thermometer well in the center, an open escape to the atmosphere at the bottom, and two threaded openings in the sides near the top and on opposite sides. One of these openings connects to the steam line. The other side leads to a *manometer,* a glass U tube partly filled with mercury and having a scale divided into inches.

Q Explain the operation of the throttling calorimeter.
A Steam from the main steam pipe passes through a nozzle having an opening about $3/100$ in. in diameter, which reduces it to approximately atmospheric pressure. Since moderately high-pressure saturated steam contains a greater number of heat units per pound (higher enthalpy) than lower-pressure saturated steam, as can be seen from Table 16-3, this excess heat superheats the steam in the calorimeter, thereby raising its

temperature above that of saturated steam at the lower pressure. The thermometer in the central well shows the temperature of the superheated steam. The manometer shows the pressure within the calorimeter.

From the observed pressure and temperature, quality of the steam can be calculated from the steam tables by the following formula:

$$q = \frac{H + 0.47(t_s - t) - h}{L}$$

where q = quality of steam

H = total heat in 1 lb steam (enthalpy of saturated vapor) at calorimeter pressure. (To get calorimeter pressure first add the barometer reading to the manometer reading. Multiply this by 0.49 to get calorimeter pressure in psia.)

0.47 = specific heat of superheated steam, atmospheric pressure

t_s = temperature of superheated steam in the calorimeter

t = temperature of saturated steam at calorimeter pressure (from steam tables)

L = Latent heat (enthalpy of vaporization) of high-pressure steam in main

h = sensible heat (enthalpy of saturated liquid) of high-pressure steam in steam main

Q Describe a separating calorimeter.

A The separating calorimeter separates water from steam, and collects it in a separate chamber where its exact weight may be determined. In Fig. 16-5 the main body of the calorimeter is a cast-iron double shell. The outer space steam-jackets the inner cylinder to prevent heat loss by radiation. A water glass with a graduated scale is connected to the inner cylinder and a pressure gage is connected to the outer steam space. This

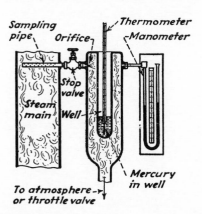

FIG. 16-4 Throttling calorimeter.

gage is graduated to show steam pressure in the outer space, and also weight of steam in pounds passing through the calorimeter orifice in a given time.

Q Describe the operation of the separating calorimeter.

A When the steam is admitted through the valve at the top of the calorimeter, it flows downward into the perforated cup where its direction of flow is reversed. Moisture in the steam, being heavier than the steam itself, is left in the cup and drains down into the bottom of the inner chamber, where the amount can be measured on the scale. Dry steam passes upward to top of cup, enters the outer cylinder, and finally passes out through the orifice in the bottom. Weight of dry steam flowing can be read on the gage dial, or it can be found by attaching a hose to the outlet at the bottom and leading this hose into a tank of water resting on a scale. The tank's weight and contents are checked before and after the calorimeter is drained into it. The difference in weight gives the weight of the condensed steam. When weights of the moisture and dry steam are found, the percentage of moisture and quality of steam can be calculated:

Percent moisture

$$\text{in steam} = \frac{\text{weight of moisture in calorimeter}}{\text{dry-steam weight} + \text{moisture weight}} \times 100\%$$

$$\text{Quality of steam} = 100 - \text{percent moisture}$$

Q Many process industries use great quantities of hot water. It is the operating engineer who must supply the hot water, and the steam for keeping it at the correct temperature. Explain the water-heating system shown in Fig. 16-6.

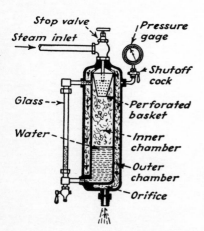

Stop valve
Steam inlet
Pressure gage
Shutoff cock
Glass
Perforated basket
Water
Inner chamber
Outer chamber
Orifice

FIG. 16-5 Separating calorimeter.

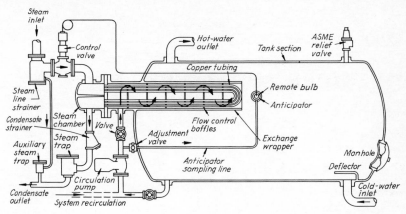

FIG. 16-6 Hot-water heater with control-flow system for providing hot process water.

A This unit keeps the entire storage volume filled with hot water except when stored water is drawn off in excess of recovery capacity. There is no delay in the heating process while strata of cold water rise to the level of the thermostatic bulb. Forced circulation assures that the temperature of the stored water is sensed continually. Whenever necessary, steam is admitted to the heat exchanger to bring stored water up to temperature.

The anticipator system responds immediately to a combination of hot-water delivery rate and incoming-cold-water temperature. It begins to admit steam before any temperature change can be sensed in the storage tank. The steam control valve is a highly sensitive pilot regulator, but overall response is always delayed by the time it takes for the valve to open, for steam to enter the heat exchanger, for steam to condense, and for its heat to be transferred to the water. The fast action of the anticipator control minimizes harmful effects of these built-in response lags.

SUGGESTED READING

Reprint sold by *Power* magazine
   Power Handbook, 64 pp.

Books
   Elonka, Stephen M., and Anthony L. Kohan: "Standard Boiler Operators' Questions and Answers," McGraw-Hill Book Company, New York, 1969.
   Elonka, Stephen M.: "Standard Plant Operators' Manual," McGraw-Hill Book Company, New York, 1975.

17

INJECTORS AND PUMPS

One of man's oldest aids, the pump today ranks second only to the electric motor as the most widely used industrial machine. Anything that will flow is pumped, even thick mud and sludge. And to meet the many demands of moving liquids, we have a confusingly large variety of available pumps. They range from tiny adjustable-displacement units to giants handling well over 100,000 gal per min. Here, we cover only the basic types used in boiler and engine rooms. The following information is the minimum that should be known by every operator of these units.

INJECTORS

Q What means are used to force water into a steam boiler against boiler pressure?
A Principal devices to feed water into steam boilers are injectors and pumps, though under certain circumstances an outside source of water supply may be suitable if pressure is high enough and a supply is constantly available. Low-pressure heating boilers, for example, are often fed from city water mains.

Q Sketch a single-tube injector and mark names of principal parts.
A Figure 17-1 is a sectional view of a single-tube lifting injector with principal parts marked.

Q Describe construction and operation of the single-tube lifting injector.
A This injector uses direct steam pressure to force water into a boiler against the same steam pressure as that which operates the injector. A brass casing contains an expanding steam nozzle, several converging and diverging tubes, which form one practically continuous passage, and a check valve.

When the steam valve is opened, steam enters the injector through the expanding steam nozzle, which converts its pressure energy into velocity. Entering the suction tube, this high-velocity steam jet creates a partial vacuum in the suction pipe. Water is then forced up the suction pipe into the injector by the unbalanced atmospheric pressure on the surface of the water supply. This water mixes with, and condenses, the steam. Some of the steam's velocity is imparted to the combined jet of condensed steam and water as it passes into the tube, and this velocity is again partly converted into pressure in the delivery tube — enough to enable the stream of water flowing through the pipe to lift the check valve and enter the boiler.

Until the jet of water is properly established in the combining and delivery tubes, steam and water escape through the small holes in the sides of the combining tube and pass out through the overflow valve. As soon as the water has gathered enough velocity and pressure to lift the boiler check valve, it passes through the combining and delivery tubes in an unbroken stream, creating a partial vacuum in the overflow chamber, which lifts the sliding washer up against its seat, thus preventing any inrush of air through the overflow opening that might break up the jet.

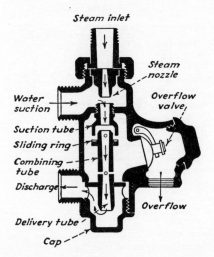

FIG. 17-1 Single-tube lifting injector.

This injector will automatically resume work after a momentary interruption in flow without having to be adjusted by the operator.

Various pressure and velocity changes that occur in the injector depend on the design of nozzles and tubes. The diverging steam nozzle allows the entering steam to expand with a consequent fall in pressure and increase in velocity. The converging suction and combining tubes increase the velocity of the water jet, and the expanding delivery tube decreases the velocity of the water jet and increases the pressure. Note that an expanding nozzle *increases steam* velocity but *decreases water* velocity, because steam can expand indefinitely while water cannot expand much.

Q Sketch a double-tube injector and name the principal parts.

A Figure 17-2 shows a sectional view of a double-tube injector with names of parts.

Q Describe construction and operation of a double-tube injector.

A It handles hotter water than the single-tube injector and works with a greater lift. It also operates with lower steam pressures and against higher boiler pressures. In operation, one tube lifts the water from the source of supply and delivers it to the other tube, which forces it into the boiler.

Capacity of the injector, Fig. 17-2, can be adjusted by the regulating valve admitting steam to the lifting tube. The lever handle operates the main steam valve, forcer steam valve, and main overflow valve. The main steam valve and forcer steam valve are operated by the same valve stem,

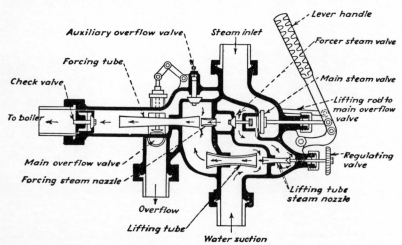

FIG. 17-2 Double-tube injector works against boiler pressure.

their construction being such that the main steam valve can be opened a certain amount before the forcer steam valve opens.

To start the injector, assuming the regulating valve is properly adjusted, pull the lever handle to the right, just enough to open the main steam valve. This admits steam to the lifting tube steam nozzle through the regulating valve. The entering high-velocity steam jet creates a partial vacuum in the suction line and lifts water from the supply source. The combined steam-and-water jet issuing from the lifting tube fills the injector body and escapes through the auxiliary and main overflow valves.

Pulling the lever handle farther to the right opens the forcer steam valve and closes the main overflow valve. Steam from the forcing steam nozzle then forces the water being supplied by the lifting tube into the forcing tube, from which it discharges with sufficient velocity to lift the boiler check valve and enter the boiler. As the main overflow valve closes, the pressure equalizes on both sides of the auxiliary overflow valve, which closes automatically.

To stop the injector, pull the lever handle to the left, closing both main steam valve and forcer steam valve.

Q What is meant by *lift* of an injector?
A It is the vertical distance from the injector to the surface of the water supply, when the injector is located *above* the supply source. Practically all injectors are of the lifting type, but it is best to place an injector as close to the water-supply surface as possible because a high-suction lift cuts down the capacity. The limit of practical suction lift at sea level is not much over 20 ft, and it is less at higher altitudes because of the lower atmospheric pressure.

Q Explain why an injector can lift water from a lower level and why it will not lift very hot water.
A An injector lifts water because the high-velocity steam jet entering the injector exhausts air from the suction pipe and creates a partial vacuum in the pipe and injector. Then the unbalanced atmospheric pressure on the surface of the water supply forces the water up the pipe into the injector.

If the feedwater is very hot, the following things happen:

1. Entering steam momentarily creates a partial vacuum by exhausting the air from the suction pipe, but continuance of this partial vacuum depends upon the incoming feedwater's condensing the steam, and this does not occur at all if the feedwater is too hot.

2. The injector becomes *steambound*. Creation of a partial vacuum in the suction pipe may cause water in the pipe to evaporate into steam, as lowering the pressure also lowers the boiling point. If this happens, the suction pipe fills with steam and no water enters the injector.

Q What are the advantages and disadvantages of an injector as a boiler-feeding device?
A Advantages are:
1. It is small and very compact.
2. Its first cost is low.
3. Upkeep and repair costs are low.
4. It has practically no moving parts.
5. Of very simple construction, it does not readily get out of order.
6. It requires no lubrication.
7. It is easily operated.
8. It has high *thermal* efficiency. This means that practically all heat in the steam used to operate the injector is also used in heating the feedwater. The only heat loss is the small amount of heat radiated from the body of injector. (But note that this "efficient" heating of feed water by live steam makes it impossible to take full advantage of otherwise wasted exhaust steam to heat the feedwater.)
Disadvantages are:
1. It cannot handle very hot water.
2. Because of limited delivery range, its operation is usually intermittent unless the boiler is carrying a steady load and injector delivery can be regulated to keep a steady water level.
3. With varying load, the injector must be started and stopped frequently, and there is danger that water level may drop below the safe minimum level when the injector is not in operation.
4. It does not operate efficiently on superheated steam.
5. It has low *mechanical* efficiency compared to a pump, its steam consumption being greater for the same amount of work done in forcing water into the boiler.

Q What are some of the common reasons for failure of an injector to work properly and efficiently?
A Failure of inefficient operation may be due to (1) feedwater too hot; (2) suction lift too high; (3) leaks in suction pipe destroying vacuum; (4) steam pressure too low; (5) wet steam preventing proper condensation of steam jet; (6) nozzles and tubes scaled up or plugged with dirt; (7) suction hose collapsed internally, if hose is used; (8) obstruction in suction or discharge line; (9) worn-out injector parts; (10) leaky boiler check valve allowing steam to blow back into injector discharge line.

Q If an injector does not work properly, what should you do to find and remedy the trouble?
A Repair leaks in suction pipe or run a new suction line. Take injector apart and examine nozzles and tubes to make sure that passages are clear and parts not badly worn. If tubes are scaled, they may be cleaned by

soaking for several hours in a dilute solution of muriatic acid, about 1 part acid to 10 parts water. Remove any iron or steel parts before placing the injector in the acid. If any parts are worn out, replace them with new parts.

Make sure that suction and delivery are free from obstruction, and that boiler check valve is in good working order. Wet steam may be caused by the steam connection to the injector being taken off at a point where considerable water comes over with the steam, in which case it may be remedied by changing the connection to a more suitable location. If water is too hot, use colder water. If lift is too high, lower injector or raise level of water supply.

Q Sketch an ejector and explain its operation.

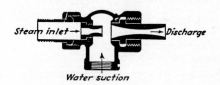

FIG. 17-3 Ejector is simple pump.

A An ejector operates on the same principle as an injector, but its construction is much simpler, and it does not discharge against a very high pressure. It is not suited to boiler feeding. It can, however, be used for many purposes where space is limited and only small quantities of water have to be handled. The type in Fig. 17-3 operates on either steam or compressed air, though steam is preferable. Steam consumption is quite high, so it should be used only where it operates intermittently and for short periods.

In operation, steam or compressed air issues at a very high velocity from the expansion nozzle and creates a partial vacuum in the suction line. Water rises in the suction pipe and when it meets the steam or air jet, it is forced into the discharge line.

Q Has a pump any advantages over an injector as a means of feeding boilers?

A A pump has several advantages over an injector. Most important are:

1. It can be adjusted to feed continuously at steady or varying rates and does not have to start and stop at frequent intervals.

2. It uses much less steam.

3. It handles hotter water.

4. It has a wider range capacity.

5. It permits the economical use of exhaust steam or bled steam for feed-water heating.

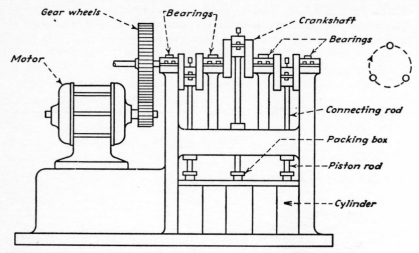

FIG. 17-4 Vertical, triplex reciprocating pump can be used for high pressures.

Because of these advantages, pumps are installed in all large steam-boiler plants and in most small plants as well, injectors being pretty well confined to very small stationary plants and portable and locomotive boilers.

Q How are pumps classified?

A A general classification is:

1. RECIPROCATING PUMPS (piston and plunger): *power driven* (simplex, duplex, and triplex); *direct acting, steam driven* (simplex and duplex).

2. CENTRIFUGAL PUMPS (steam turbine, electric motor, or belt driven): *single-stage* or *multistage*.

3. ROTARY PUMPS (turbine, motor, or belt driven): with impellers in the form of gears, lobes that mesh together like gears, or sliding or swinging vanes.

Q Describe a power-driven pump.

A Strictly speaking, all pumps are power-driven in some fashion, but the term *power pump* is usually limited to a pump of the reciprocating type that is driven by a belt or gears from an engine, motor, or line shaft. The pump may have one, two, three, or more cylinders, and the cylinders may be single or double acting. The *triplex* is a very common form of power pump. It has three cylinders set side by side, with their plungers connected to a three-throw crankshaft. The cranks are set 120° apart to ensure a fairly steady flow of water.

Plungers are usually *single acting;* that is, the upward stroke is a suction stroke and the downward stroke is a discharge stroke, the upper end of the cylinder being open.

Figure 17-4 shows a vertical gear-driven triplex power pump, often used as a boiler feed pump.

RECIPROCATING PUMPS

Q Explain a direct-acting duplex steam pump.
A A sectional sketch of one side of a direct-acting duplex steam pump is shown in Fig. 17-5. It is called a *duplex* pump because it has two steam and two water cylinders, placed side by side.

Q Describe the action in the water end of a duplex direct-acting steam pump.
A For a pump set above the surface of the water supply, the action is as follows: When the pump is started, the movement of the pistons to and fro exhausts the air from the suction pipe and water cylinders. This creates a partial vacuum in the suction line, and the unbalanced atmospheric pressure on the surface of the water supply forces water up the suction pipe into the pump cylinders. Water passes up through the suction valves into the cylinder on one stroke, and is forced up through the discharge valves on the return stroke, as the pressure of the piston on the water closes the suction valves and opens the discharge valves.

In Fig. 17-5, atmospheric pressure is forcing the water into the left-hand side of the cylinder. The piston is forcing the water on the right to flow out through the discharge valves. Since the pump is *double acting*, this process reverses on the return stroke.

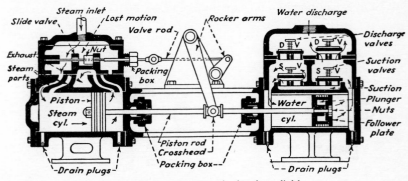

FIG. 17-5 Duplex direct-acting steam pump is simple, reliable.

Q What is the difference between a plunger and a piston?

A Though both serve the same purpose, they differ in construction. A plunger is a long cylinder, closed at one or both ends, depending upon the type of pump. In some single-acting power pumps, the outside end of the plunger is open, and the plunger is directly connected to the crank shaft by a connecting rod. In other types of power-plunger pumps, the plunger is closed at both ends and is driven by a short rod connected to a cross-head. In direct-connected steam-driven plunger pumps, the plunger is connected to the same rod as the steam piston.

In all types of plunger pumps, the plunger does not fit closely in the cylinder but is kept airtight and watertight by passing through an *outside-packed* packing box, as shown in Fig. 17-6*b*.

A water piston is a short piston similar in construction to a steam piston. It fits closely inside the pump cylinder and is *inside packed;* that is, it is made watertight by means of rings of square packing fitted into a groove around the circumference of the piston and held in place by a follower plate and a nut on the end of the piston rod. The packing commonly used is made from layers of cotton fabric cemented together with rubber. Sometimes harder materials are used to give longer wear or when the pumps are to be used for pumping liquids that would eat away and destroy softer packing materials. A water piston packed with square rings is shown in Fig. 17-6*a*.

Q Explain the action in the steam end of a direct-acting duplex pump.

A In Fig. 17-5, two simple steam-engine cylinders are set side by side, the slide valve in each cylinder steam chest being operated by the cross-head on the piston rod of the opposite cylinder, through an arrangement of rods and rocker arms.

Steam valves are the simple *D* type. They slide to and fro over the steam ports, admitting steam alternately to each end of the steam cylinder and exhausting it to the atmosphere when the piston stroke is almost

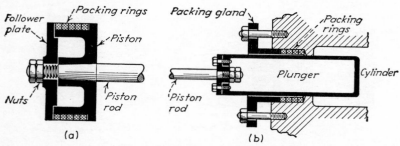

FIG. 17-6 (*a*) Inside packed piston; (*b*) outside packed plunger.

completed. There are two steam ports for each end of the cylinder. Steam enters the cylinder through the outer port and passes to exhaust through the inner port, after it has done work on the piston. These separate ports cushion the piston at the end of its stroke and prevent its striking the cylinder heads. As the piston nears the end of its stroke it covers the exhaust port, preventing further escape of steam to exhaust. The steam thus trapped between piston and cylinder head brings the piston to rest and reverses its motion without shock.

Some pumps have hand-operated cushion valves that open or close small passages connecting the steam and exhaust passages at each end of the cylinder, and so regulate the degree of cushioning. These valves should be closed at light loads to give the maximum cushioning effect and shortest possible piston stroke. At heavy loads, the cushion valves should be wide open to give minimum cushioning effect and longest possible piston stroke.

Because each steam valve is operated by the action of the piston rod belonging to the other steam cylinder, the pistons move in opposite directions most of the time. When one piston is momentarily stopped at the end of its stroke, the other piston is in motion, thus ensuring a fairly continuous water flow in the discharge line. This is one point where the duplex pump has an advantage over the simplex type of direct-acting pump. Another is its comparative simplicity of construction and operation.

Q Describe a simplex direct-acting steam pump.
A The action in the water end of this pump, Fig. 17-7, is similar to the action in the water end of the duplex pump, but the water valves are placed at the side instead of on top of the water cylinder, as in the duplex pump. This makes the valves readily accessible without having to disconnect any pipes or remove more than one cover.

The main difference between the simplex and duplex steam pumps lies in the method of operating the steam valves. The entire valve mechanism is contained within the pump itself, and there are no outside moving parts such as valve rods and rocker arms. This pump cannot *short-stroke*, because the main piston must complete its travel before the main steam valve can reverse.

Q Describe the steam-valve operation in the simplex pump shown in Fig. 17-7.
A The main steam piston is driven by steam admitted alternately to each side of the piston by a main slide valve that also allows steam to pass to exhaust when it has done its work in moving the piston. This main slide valve is moved back and forth by a plunger or close-fitting piston valve with hollow ends. These ends are open to the steam chest and are always

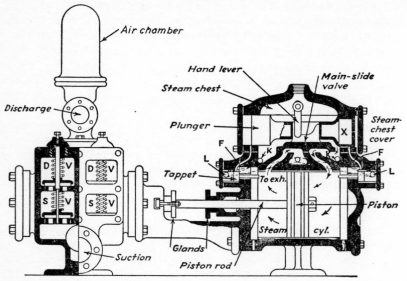

FIG. 17-7 Simplex direct-acting steam pump with air chamber.

filled with live steam, which flows through tiny holes marked x in the ends, thus filling the spaces between valve ends and valve-chest covers with live steam also.

Since pressure on the piston-valve ends is equal, it is ordinarily balanced and motionless. When the main piston travels to the end of its stroke, it strikes the projecting spindle of a small tappet and pushes the tappet back. This places passage F in communication with the exhaust through the passage shown dotted and allows live steam in the space at the end of the piston valve to escape to exhaust. Pressure of the live steam on that end being now relieved, the piston valve is forced over by the live-steam pressure on the other end. As it moves it carries the main slide valve with it, thus reversing the motion of the main piston by admitting steam to the opposite end of the cylinder.

As the piston valve moves over, it shuts off port F and cushions itself on the steam trapped between end of valve and steam-chest cover. Except when pushed in by the main piston at the end of its travel, the tappets are kept closed by constant live-steam pressure on their larger ends, conveyed through the passages KL from the steam chest.

The short lever shown in the center of the large piston valve can be operated by a similar lever on the outside of the steam chest. It is used to start the pump if it happens to stop with the piston valve and slide valve exactly on center.

These valves do not have to be set, since they are operated entirely by steam pressure and are not attached to any moving pump part. They must, however, be kept well lubricated because of their large area of rubbing surface.

Q Sketch some type of vertical steam-driven reciprocating steam pump, and describe the main points in its construction.

A Figure 17-8 is a sketch of a vertical simplex steam pump, showing the steam and water cylinders in section and a section through one set of water-suction and discharge valves, which are connected to the upper part of the water cylinder. Single suction and discharge valves are shown in the sketch, but some pumps have a number of small valves instead of one large valve. These valves are in a separate chamber attached to the water cylinder, and are readily accessible for inspection or repair without having to take any other sections of the pump apart. The area of the delivery valve, or valves, is usually less than the area of the suction valve, or valves, so as to increase the velocity of flow in the discharge line. This also has a tendency to improve the operation of the pump when it is pumping very hot water. Like most other simplex pumps, the steam

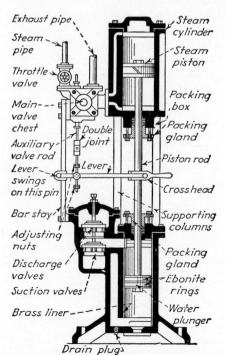

FIG. 17-8 Vertical simplex steam pump.

valves are so constructed that the piston must take a full stroke before it can reverse.

The vertical type of reciprocating pump takes up less floor space than the horizontal type, and there is less wear on the cylinders and pistons, as the weight of the pistons does not rest on the cylinder walls.

Q Explain the valve action in the steam end of the vertical simplex pump shown in Fig. 17-8.

A The steam chest containing the main and auxiliary steam valves is placed with its axis at right angles to the center line of the pump cylinders, so that the main steam valve moves to and fro horizontally. This arrangement is to prevent the possibility of the weight of the valve causing it to drop down and make the pump stroke short or uneven, as it might do if the travel were vertical instead of horizontal. The main valve is cylindrical in shape with a flat machined on the back to form a seat for the auxiliary valve. The auxiliary valve is a small flat slide valve which slides vertically up and down on the back of the main valve and is operated by a crosshead on the pump-piston rod.

Q How is a steam reciprocating pump controlled?

A Figure 17-9 shows a hookup for a simple manually adjusted regulator. While it controls pressure of steam to the steam chest of the pump, it is not sensitive to change in demand. If a reciprocating pump is operated with reasonably constant capacity or load, it will maintain constant discharge pressure, but only if the steam pressure to the pump's valve chest is fairly constant.

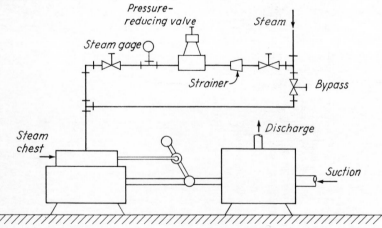

FIG. 17-9 Manually adjusted regulator controls only pressure of steam supply to pump.

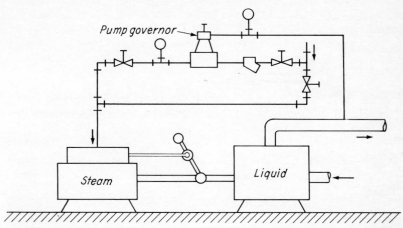

FIG. 17-10 Automatic regulation against changes in both discharge pressure of liquid and steam supply to pump.

To assure constant steam pressure, the regulator is hooked up as shown in Fig. 17-9. It is an internal-pilot-type piston-operated pressure-reducing valve. A tiny pilot in the valve's bonnet provides accuracy, while the large piston in the valve's body provides high valve lift for quick action.

Q Describe a method of controlling a steam reciprocating pump more accurately than in the previous question.

A Figure 17-10 shows the hookup for regulating the reciprocating pump against changes in the output discharge pressure and also in the steam pressure. The pump governor used is the dial-diaphragm-type regulator which is sensitive to changes in (1) steam-chest pressure and/or (2) pump-discharge pressure, to both of which it is connected. Thus a change in either pressure serves to activate the governor valve in the steam line that controls the pump's chest pressure.

SETTING VALVES

Q How would you set the steam valves on a duplex pump?

A The first step is to place the pistons in their central positions. They should be at midtravel when rocker arms are vertical, but simply moving the pistons until the arms are vertical cannot be depended upon, since the crossheads may have moved on the piston rods from their original setting. The correct procedure is as follows:

1. Push one of the pistons back until it strikes the cylinder head, Fig. 17-11.

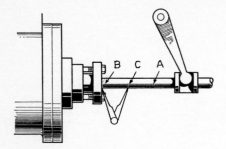

FIG. 17-11 Setting valves of du-plex reciprocating steam pump.

2. Scribe mark *A* on the piston rod at the face of the packing gland.

3. Pull the same rod out until the piston strikes the opposite cylinder head.

4. Scribe another mark *B* on the piston rod at the face of the same packing gland.

5. Find the center of the distance between these two points and mark this central point *C*.

6. Move the rod back until the central mark is at the face of the gland, and the piston will be exactly at mid-stroke. Repeat this process for the other side of the pump.

7. Set the slide valves exactly in the center of their travel. In this position they will just cover the steam ports with no overlap.

8. Divide the lost motion evenly on each side of the nut or nuts on the valve rod that moves the valves.

9. The setting is now complete, but before replacing the valve-chest cover, be sure to move (with a finger) each valve as far as it will go, in opposite directions, so that the steam ports are uncovered, otherwise the pump will not start when steam is turned on.

To summarize, valves are properly set if they just cover the steam ports, with lost motion equally divided, when the pistons are at mid travel. The term *lost motion* refers to the space between the lugs on the back of the valve and the nuts on the valve rod that moves the valve. This slack is necessary to let the piston almost complete its stroke before the crosshead, rods, and rocker arms move the valve. If no lost motion were provided, the pistons would either remain stationary or make a very short stroke.

Q What means are used to automatically regulate boiler feed-pump discharge to suit varying boiler loads?

A *Pump governors* are used to regulate the speed of boiler feed pumps of the reciprocating type. These act to throttle the steam supply to the pump when pressure builds up in the discharge line through the partial closing of the boiler feed valves.

Centrifugal boiler feed pumps usually run at constant speed. When pressure builds up in the discharge line, it opens a relief valve through which the water flows back to the supply tank.

Q Describe a pump governor.

A Figure 17-12 shows a sectional view of a diaphragm-actuated pump governor. The steam supply to the pump steam cylinder passes through a double-seated valve so proportioned that the upward pressure on the upper valve disk balances the downward pressure on the lower valve disk. The space above the diaphragm is connected to the feed-pump water-discharge line, and any increase in the discharge pressure forces the diaphragm downward. This in turn pushes the valve stem downward and closes the steam valve. When the water-discharge pressure falls, the upward pressure of the heavy coil springs forces the diaphragm and valve stem upward and opens the steam valve. The amount of spring compression and the working limits of the governor can be regulated by adjusting nuts threaded on the valve stem.

Q What is an *air chamber,* and what duty does it perform?

A An *air chamber,* Fig. 17-7, is a hollow casting, closed at the upper end and having a flange at the bottom end for attaching it to the pump or pipe line. Air chambers are placed in the discharge line near the pump or directly on top of the discharge-valve chamber.

The air chamber steadies the flow of water, which is more or less intermittent in any reciprocating pump. The air within the air chamber is

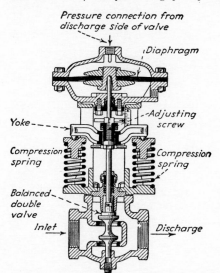

FIG. 17-12 Diaphragm type pump governor.

compressed when the discharge pressure is greatest, and it expands and keeps the water in motion when the discharge pressure is lowest. This also permits the valves to seat with less shock than would be the case if no air chamber were used. Air chambers are sometimes placed on suction as well as discharge lines to give a steadier flow of water in the suction line.

If an air chamber becomes filled with water, it should be drained, since it has no beneficial effect on the operation of the pump when in this state. Most air chambers have small drain valves to drain off water when necessary, and large air chambers usually have a gage glass to show height of water and presence of air in the chamber.

Q What is an ALLEVIATOR?

A Air chambers are of little use where water pressures are very high, since the air is forced through the pores of the iron or is absorbed by the water, but a device called an *alleviator*, Fig. 17-13, can be used for the same purpose. It consists of a cylinder, flanged at one end for attachment to the pipe line, and having a packing gland at the other end through which a close-fitting plunger can slide freely. The upper end of the

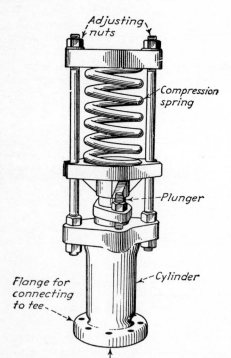

FIG.17-13 Alleviator exhausts air to atmosphere.

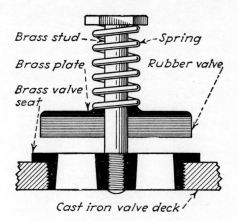

Brass stud

Spring

Brass plate

Rubber valve

Brass valve seat

FIG. 17-14 Valve assembly for liquid end of reciprocating pump.

Cast iron valve deck

plunger is attached to a heavy coil spring with means for adjusting the compression of this spring as desired. Excessive pressure in the line forces the plunger upward against the pressure of the spring, and so relieves the pressure in the pipe line. As the pipe-line pressure falls, the spring forces the plunger back down into the cylinder.

Q Sketch and describe a pump water-valve assembly as commonly used in reciprocating pumps.

A Figure 17-14 shows a complete valve assembly. The valve is a rubber disk, soft rubber being used for pumping cold water and hard rubber for hot water. The valve fits closely on a brass seat which is screwed or pressed into the valve-deck casting. The valve is held down on its seat by a brass plate and spring, and these in turn are held together by a stud screwed into the seat. The valve is opened by the water pressure on the under side lifting it against the pressure of the spring. Pump valves are also made from cast and pressed metal and various plastic materials.

Q What is meant by *priming* a pump?

A *Priming* a pump means filling the suction line and water cylinders or pump casing with water before starting the pump. The pistons or plungers of a reciprocating pump may not be able to exhaust air from, and create a partial vacuum in, the suction line if the pump is set very high above the surface of the water supply and the valves and pistons are badly worn. In that case it may be possible to start the pump and keep it in operation by priming, as the water forms a seal. However, this should not be necessary with a reciprocating pump in good shape and with a moderate suction lift.

The impellers of centrifugal pumps do not have an airtight fit in the pump casing, and their rotation cannot reduce the pressure in the suction line much below atmospheric pressure. It is therefore always advisable to place a centrifugal pump below, or as close as possible to, the surface of the water supply. When the pump is actually lower than the water supply, it will start without any trouble, as the water will flow into the pump by gravity, but when the pump has to lift the water, it must be primed before it will start.

If a check valve is placed on the end of the suction pipe, the pump may be primed by filling the suction line and pump casing with water from some other source of water supply, such as city water mains or the pump discharge line if it should happen to be standing full of water above the pump. When the priming water is turned on, the cocks on top of the pump casing should be opened to let the air escape to indicate when the casing is full of water. A steam- or air-driven ejector may also be used to exhaust air from the pump and suction line and cause the water to rise up into the pump. Various automatic arrangements are used to prime and start centrifugal pumps that operate intermittently and with a suction lift. Most of these use float-controlled switches and electrically driven auxiliary priming pumps or valves on water-pressure lines.

Q Sketch and describe a *foot valve.*
A This is a check valve that will open and allow water to pass freely up the suction pipe to the pump, but will close and hold the suction line full of water when the pump stops, thus making it much easier for the pump to pick up its water when it is again started. Figure 17-15 shows a sectional view of a foot valve fitted with a strainer to prevent any large pieces of debris, such as chips of wood, from passing up the suction pipe into the pump. The valves are brass disks, faced with leather or rubber.

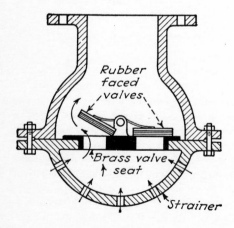

FIG. 17-15 Foot-valve assembly.

Q How would you repack the pistons and plungers in inside- and outside-packed reciprocating pumps?

A To repack the piston of an inside-packed pump, first remove the cylinder cover. Unscrew the large nut on the end of the piston rod. Pull out the follower plate. Remove the old packing, using a packing hook to pull it out. Cut new rings to the correct length, making them a trifle short so as to allow for expansion. Put new rings in place, being careful to *break joint*, that is, not to place joints of rings opposite each other so that water may leak through. Replace follower plate, and screw large nut on end of piston rod. Replace cylinder cover, making a new gasket if necessary. Take care not to put too many rings of packing on the piston; tightening up the follower plate plus the swelling of the packing by water may cause the piston to bind. The piston rod is packed with any good grade of rod packing, preferably one impregnated with a lubricant such as graphite.

Outside-packed plungers can be repacked without taking the pump apart, it being necessary only to loosen and pull back the packing glands, pull out the old packing, and insert new rings. A softer packing can be used than that used for inside-packed pumps, since the glands can be tightened up from time to time without the pump being shut down. Repacking also is a much quicker and easier process than with the inside-packed pump.

CENTRIFUGAL PUMPS

Q Explain the operation principle of a centrifugal pump.

A In a centrifugal pump, one or more revolving wheels, called *impellers,* are attached firmly to a central shaft and surrounded by a stationary casing. As these impellers revolve, water enters at the center from the suction line and is thrown outward by centrifugal force. Water leaves the impeller rim at high velocity, and this velocity is converted into pressure either in the surrounding casing or by diffusion rings surrounding the impeller. In a single-suction pump, water enters one side of the impeller only. In a double-suction pump, water enters the impeller on both sides.

If a centrifugal pump contains only one impeller, it is a *single-stage* pump. If it contains two impellers, it is a *two-stage pump,* and so on. Multistage pumps are constructed with as many as 10 stages. Single-stage pumps are used mainly for pumping against low heads and moderate pressures, although improvements in design have greatly increased the range of this pump. In the multistage pump, delivery from the first stage passes to the suction of the second stage and is delivered from the second-stage impeller at an increased pressure. In this way pressure is boosted in succeeding stages so that the more stages used, the higher will be the final pressure. In practice, the maximum number of stages that can be used is limited by mechanical difficulties of construction.

Q Describe construction of the impellers in centrifugal pumps.
A Figure 17-16 shows two common types, the *closed* and the *open* impeller. The closed impeller has a plate or *shroud* on each side of the vanes, whereas the vanes of the open impeller are not in any way closed in. Side plates of the closed impeller direct the flow and thus make it more efficient than the open impeller for pumping clean water. But the open unit is better for handling thicker liquids such as heavy oil, sewage, or water containing a great deal of dirt or sand.

Q Explain how the turbine and volute principles are used in centrifugal-pump design.
A In the turbine centrifugal pump, the impeller is surrounded by a stationary diffusion ring containing passages of gradually increasing cross-sectional area. Water leaves the impeller rim at a high velocity which is converted into pressure as the water passes through the diffusion ring. The casing surrounding the diffusion ring is usually circular in shape and of constant cross-sectional area.

In the volute pump no diffusion rings are used, but the casing is volute in shape—that is, of gradually increasing cross-sectional area—so that as the water flows in the casing its velocity is reduced and its pressure increased.

Q Describe a centrifugal pump using the turbine principle.
A Figure 17-17 is a sectional view of a three-stage centrifugal pump having two turbine stages and one volute stage. The liquid from the last impeller flows directly to the discharge, and the use of a final volute stage simplifies construction to quite an extent.

As the impellers are all single-suction, the end thrust is considerable. It

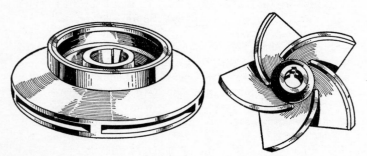

FIG. 17-16 Closed impeller (left) and open impeller (right).

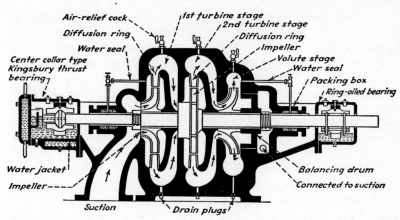

FIG. 17-17 Centrifugal pump with two turbine stages and one volute stage.

is opposed by water pressure acting upon the balance drum in the last stage, and any excess end thrust acting in either direction is taken care of by a center-collar-type Kingsbury thrust bearing. The other bearing is the ring-oiled type.

Q Describe a centrifugal pump using the volute principle.

A Figure 17-18 shows a sectional view of a single-stage double-suction

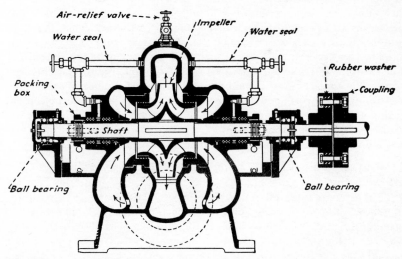

FIG. 17-18 Single-stage volute-type centrifugal pump and coupling.

volute pump. In this pump no diffusion rings are used, but the velocity of the water leaving the impeller is converted into pressure in the volute casing. The hydraulic balance is practically perfect in this pump because of the double suction and direct flow to the discharge, so end thrust is negligible. The shaft is supported by ball bearings, one being fixed and the other free to move a little endwise and thus take care of any slight changes in shaft length that result from expansion or contraction.

Use of volute pumps was formerly confined very much to low-pressure work, but volute-type pumps are now made to work against high pressures, in both single-stage and multistage patterns. The volute pump is of simpler construction than the turbine pump and has fewer parts.

Q What methods are used to balance end thrust in centrifugal pumps?
A End thrust may be taken care of by balance plates, ball bearings, multicollar thrust bearings, and special forms of single-collar thrust bearings. The balance plate or drum, as shown in Fig. 17-17, is keyed firmly to the shaft, and its area is such that the pressure of water acting upon it balances the end thrust in the opposite direction.

A ball-thrust bearing, suitable for light or medium end-thrust loads, is shown in Fig. 17-19a. For heavier loads, a multicollar bearing may be used, Fig. 17-19b. A number of collars turned from the solid shaft fit into recesses in the top and bottom halves of the bearing. The multicollar bearing has been largely superseded by single-collar bearings where the thrust is taken up by a number of segments that are flat on the side next to the thrust collar but pivoted at the back so that they can tilt in any direction. A constant stream of oil is supplied to the bearing, and when the collar presses against the bearing surface of the segments, they assume the position shown in Fig. 17-19c, allowing a wedge-shaped stream of oil to pass between the bearing surface of each segment and the collar. The thrust is thus actually supported by a film of oil, and there is little or no metallic contact between the bearing surfaces. If the bearing is meant to support end thrust in either direction, the collar is placed in the center of the bearing with a ring of bearing segments on each side. This type of thrust bearing can support a much greater load per square inch of bearing surface than the multicollar type of thrust bearing.

Q What valves are necessary in the suction and discharge line of a centrifugal pump?
A If the pump is placed above the surface of the water supply, there should be a foot valve on the end of the suction line and some means of priming the pump before starting. If the pump is placed lower than the surface of the water supply, no foot valve is needed on the end of the suction line, but a stop valve, preferably of a straightway type, must be

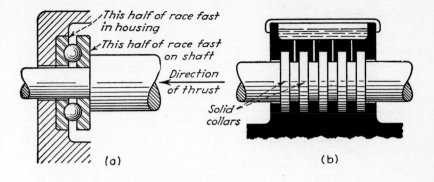

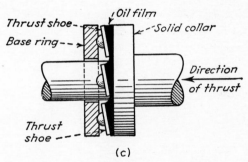

FIG. 17-19   (*a*) End-thrust ball bearing; (*b*) multicollar thrust bearing; (*c*) Kingsbury thrust bearing.

placed in the suction line at the pump to enable the flow of water through the suction to be cut off when necessary. In both cases, a check valve should be placed in the discharge line close to the pump, and beyond that, a straightway stop valve.

Q How do you start and stop a centrifugal pump?

A If the pump is above the water-supply surface, prime it first, bring it up to speed, then open the discharge valve gradually to put the load on by degrees. If the pump is below the surface of the water supply, open the suction valve, start the pump, bring it up to speed, then open the discharge valve gradually.

To stop a centrifugal pump, close the discharge valve to take off the load. Stop the pump. Close the suction valve if there is a stop valve in the suction line.

If the pump has to be primed, there should be a foot valve on the end of the suction line.

ROTARY PUMPS

Q What is a *rotary* pump?

A This is a pump in which the liquid is pumped by means of rotating elements of various shapes, contained within a closely fitted casing. Because of the practically airtight fit of the impellers in the casing, their rotation creates a partial vacuum in the suction line when the pump is started, and thus permits the pump to be set with a suction lift. Unlike

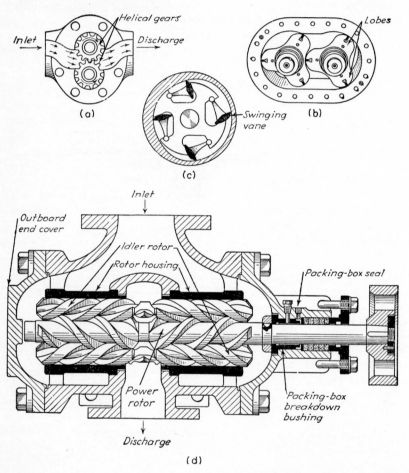

FIG. 17-20 Rotary pumps. (*a*) Gear type; (*b*) lobe type; (*c*) swinging-vane type; (*d*) screw type.

the centrifugal pump, the displacement is *positive* and entirely independent of velocity of flow and centrifugal force.

Q Describe some common types of rotary pump.

A Figure 17-20a shows the operating principle of the *gear* type of rotary pump. The pump body contains two spiral gears that mesh closely with each other and have a minimum amount of clearance between the points of the gear teeth and the casing wall. As the gears revolve, liquid is drawn in at the suction opening, carried around between the gear teeth and the pump casing, then forced into the discharge line.

The pump shown in Fig. 17-20b is somewhat similar in principle to (a), but the impellers have large *lobes* instead of teeth. Meshing gears on the ends of the impeller shafts drive the impellers and keep the lobes spaced so that they do not actually touch each other. Close contact between the points of the vanes and between the sides of the casing and the lobes is secured by flat metal strips pressed outward by springs.

In Fig. 17-20c the drive shaft and rotor are eccentric with the casing bore, and the pumping is done by swinging vanes that push back into recesses in the rotor during one part of the revolution and swing outward during the rest of the revolution to carry the liquid from the suction opening to the discharge.

Figure 17-20d shows a *screw* pump in which the liquid is drawn in at the center and forced outward to the discharge along the spiral grooves in the rotating impellers.

Reciprocating and centrifugal types of pumps are usually preferred for water pumping, but compact form, simplicity of construction, and positive displacement make the rotary pump very suitable for pumping heavy oils and other thick or viscous liquids possessing some lubricating property.

Q What is an *air lift* and how does it operate?

A An *air lift* is a device for raising water from wells by compressed air, no moving mechanical parts being used. As shown in Fig. 17-21, a large pipe is placed in the well with the lower end submerged. Compressed air is led into the bottom of the large pipe through a small air line, run either inside the larger pipe, as shown in sketch, or on the outside and curved up into the larger pipe at the bottom.

As the compressed air escapes into the water at the bottom of the suction pipe, it produces a mixture of air bubbles and water which is lighter than the well water on the outside of the suction pipe. This column of air bubbles and water is forced upward by the pressure of the heavier column of water outside. A supply of compressed air is necessary to operate the air lift.

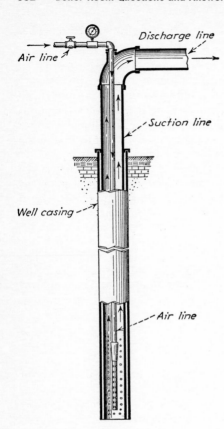

FIG. 17-21 Air lift details.

Rate of discharge depends upon *lift* and *submergence*. Lift is the vertical distance from the surface of the water in the well to the point of discharge. Submergence is the vertical distance from the point at which the compressed air enters the suction pipe to the surface of the water in the well. The depth of water in the well is, of course, lowered to some extent when the air lift is in operation, so the running submergence varies in practice between 35 and 75 percent of the total vertical distance from point of air entry to point of water discharge.

Q What is an *air pump*?

A An *air pump* (also called a *vacuum pump*) is a pump used to exhaust air from an enclosed vessel or system such as a steam-engine condenser or a heating system. Reciprocating and centrifugal pumps of special design are used for this purpose, and also steam jets.

SUGGESTED READING _____

REPRINTS SOLD BY *Power* MAGAZINE
Balancing Rotating Equipment, 24 pp.
Bearings and Lubrication, 32 pp.
Pumps, 32 pp.

BOOKS
Beard, Chester S: "Final Control Elements," Rimbach Publications Division, Chilton Company, Philadelphia, Pa., 1969.
Elonka, Stephen M.: "Standard Plant Operators' Manual," 2d ed., McGraw-Hill Book Company, New York, 1975.
Elonka, Stephen M., and Joseph F. Robinson: "Standard Plant Operators' Questions and Answers," vols. 1 and 2, McGraw-Hill Book Company, New York, 1959.

18

PUMPING THEORY AND PROBLEMS

Humans live on the floor of an ocean of air which weighs 14.7 psi. And because air has weight, humans use it to lift liquids by reducing the pressure between the liquid level and the suction side of a pump. There are many things that must be learned about pumps in order to solve each problem that comes up when working with them. Here, we round out the theory begun in the preceding chapter and cover some calculations the operating engineer should know.

PUMPING THEORY

Q What is meant by *static suction lift* in connection with pumps and injectors?

A *Static suction lift* is the vertical distance in feet, Fig. 18-1, from the surface of the water supply to the intake of the injector or center line of the pump suction, when the pump or injector is placed above the source of water supply. Always slope lone suction lines upwards toward the pump as illustrated, to avoid air pockets forming at the high point and thus reducing pump capacity.

Q Where a suction line loop for a centrifugal pump is unavoidable when going through a wall, or some object is in the way, illustrate how you would solve the problem.

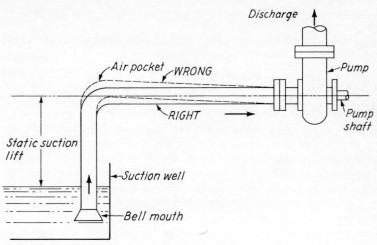

FIG. 18-1 Right and wrong methods of installing long suction line.

A Figure 18-2 shows the *right* method, dropping instead of raising the loop. Even when pumping from a pressure tank as in the figure, the pump would not deliver the expected flow with an upward loop, in which air and other noncondensable gases liberated from the water will cause a centrifugal pump to lose capacity.

Q What is *dynamic suction lift?*
A This is static suction lift plus friction and velocity losses in the suction line.

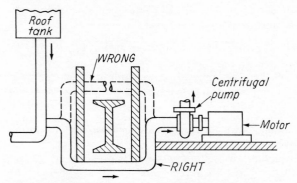

FIG. 18-2 Right and wrong methods of looping past obstruction.

Q What is the maximum theoretical static suction lift of a pump or injector?

A The maximum theoretical static suction lift depends upon the atmospheric pressure. At sea level (where the atmospheric pressure is approximately 14.7 psia), if a long length of pipe were plugged at one end, filled with water, inverted without spilling any of the water, and the open end submerged in a tank of water, the atmospheric pressure on the surface of the water in the tank would support a column of water in the pipe 34 ft high. This, then, would be the maximum theoretical static suction lift of a pump or injector at sea level, if the steam jet or pump plungers could create a perfect vacuum in the suction pipe.

Atmospheric pressure decreases as we ascend above sea level, and so, also, does the maximum theoretical static suction lift. Dividing the atmospheric pressure at sea level by the height of the column of water that can be supported by the atmospheric pressure (14.7 ÷ 34) gives the constant number 0.434. Then to find the maximum theoretical static suction lift for any given atmospheric pressure, divide the atmospheric pressure in psia by 0.434, and the answer will be the maximum theoretical static suction lift in feet.

Q How is atmospheric pressure usually measured?

A It is usually measured by the height in inches of a column of mercury in a glass tube. A mercury barometer for indicating atmospheric pressure can be made by taking a long glass tube closed at one end, filling it with mercury, inverting the tube without spilling any mercury, and immersing the open end in a dish containing mercury. The atmospheric pressure on the surface of the mercury in the open dish holds the column of mercury up in the tube just as a column of water is held in the suction line of a pump or injector by atmospheric pressure, but as mercury is heavier than water, the column is not so high. The height of the mercury column in a mercury barometer is around 30 in. at sea level.

Q How can atmospheric pressure as measured in inches of mercury be converted into pounds pressure per square inch?

A Since 1 cu in. of mercury weighs 0.49 lb, we multiply the reading in inches of mercury by 0.49 in order to convert it into pounds pressure per square inch.

Q What would be the maximum theoretical static suction lift of a pump when the mercury barometer read 26.4 in.?

A Multiply inches of mercury by 0.49 to get pounds pressure per square inch, then divide by 0.434 to get lift in feet, thus:

$$26.4 \times 0.49 = 12.936 \text{ psi}$$
$$\text{Suction lift} = 12.936 \div 0.434 = 29.8 \text{ ft} \quad \textit{Ans.}$$

Q Why is the practical suction lift of a pump or injector always less than the theoretical suction lift?

A The maximum theoretical suction lift cannot be attained with the ordinary pump or injector because of the frictional resistance to the flow of water offered by the piping and fittings, the pressure used up in imparting velocity to the water, and the practical impossibility of securing a perfect vacuum in the suction line.

The temperature of the water also has a bearing on the practical suction lift. If the water is very hot, it may vaporize under the reduced suction pressure, filling the pump and suction line with steam and thus preventing water from rising in the suction line by destroying the partial vacuum. Even moderately warm water reduces the possible suction lift by the head equivalent to the vapor pressure (see steam tables in Chap. 16, Tables 16-3 and 16-4). Thus for water at 102°F abs, steam or vapor pressure is 1 psi. This will reduce head by $1 \div 0.434 = 2.3$ ft. When pumping cold water, the practical suction lift is never more than about 80 percent of the maximum theoretical suction lift.

Q What is meant by *head* in connection with pumping?

A The term *head* is commonly used to express difference of elevation in feet or the equivalent in psi, but in pumping there are other factors to be considered. The vertical distance in feet from the center line of the pump to the point of free discharge (or the free surface of the liquid in the discharge well) is the *static discharge head*. The static discharge head plus the equivalent head required to overcome friction and impart velocity to the water in the discharge line is the *dynamic discharge head*.

The *total dynamic head* against which a pump operates is the sum of the dynamic suction lift and the dynamic discharge head.

Q What is the relationship between head in feet and pounds pressure per square inch?

A 1 lb pressure per sq in. $= 1 \div 0.434 = 2.3$ ft of head.
 1 ft of head $= 1 \div 2.3 = 0.434$ psi   *Ans.*

Q Tabulate some common data used in pumping calculations.

A Some of the principal constants used in pumping calculations and their equivalent values in common units of weight and volume are as follows:

 1 lb pressure per sq in. $= 2.3$ ft of head
 1 ft of head $= 0.434$ psi
 1 cu ft of water weighs 62.4 lb
 1 cu ft of water contains 7.48 U.S. gal
 1 cu ft of water contains 6.24 imperial (imp.) gal
 1 U.S. gal weighs 8.33 lb

1 imp. gal weighs 10 lb
1 U.S. gal contains 231 cu in.
1 imp. gal contains 277 cu in.
1 cu ft of hot water weighs approximately 60 lb
1 boiler hp is equivalent to the evaporation of 34½ lb of water from water at 212°F into saturated steam at 212°F.
1 ft-lb is the mechanical work done when a weight of 1 lb is lifted through a height of 1 ft, or work done when a force of 1 lb is exerted through a distance of 1 ft. It is the unit of mechanical work.

In addition to the above quantities, the following formulas relating to circles and cylinders are used in many pumping calculations;

The area of a circle is 78.5 percent of the area of the square it fits into. That is the diameter squared and then multiplied by 0.785, or

$$A = D^2 \times 0.785$$

The diameter of a circle is equal to the square root of the quotient of the area divided by 0.785, or

$$D = \sqrt{A \div 0.785}$$

The volume of a cylinder is equal to the area of the end multiplied by the length, or

$$V = A \times L$$

The values given have been carried to three significant figures only. Practically all pumping calculations are based on very approximate data and observations. Using more exact values would therefore mean a great deal more arithmetical work with no real advantage in the accuracy of the final result. For the same reason it is a mere waste of time to take approximate measurements and work an answer out to some such figure as 24,689.764 when all we can be reasonably sure of is that the quantity is somewhere around 24,700.

CALCULATIONS

Q How do we calculate the *useful work* done by a pump?
A *Useful work* done is the product of the water pumped, in pounds, multiplied by the height in feet to which the water is raised. This gives foot-pounds of work done.

Q How do we calculate the *useful horsepower* of a pump?
A The *useful horsepower* is found by dividing the useful work, in foot-pounds, done *in one minute* by the unit of horsepower (33,000 ft-lb of work done in 1 min).

Q What is the *mechanical efficiency* of a steam pump, and how is it calculated?

A The mechanical efficiency of a steam pump is the percentage of the power developed in the steam cylinder that is delivered as useful work in raising water against a certain head or pressure. In the case of any other type of pump, it is the percentage of the power put into the pump, which is delivered as useful work. Useful power delivered is always less than power supplied to the pump, because some power is used up in overcoming friction of the pump parts.

Stated as a formula,

$$\text{Mechanical efficiency} = \frac{\text{power output of pump}}{\text{power input of pump}}$$

Q What is the horsepower developed in a simple steam-pump cylinder of 8-in. diameter and 12-in. stroke, with the pump making 50 working strokes per minute, and with steam pressure 100 psi gage?

A Use the common formula for horsepower of steam engines; thus,

$$ihp = \frac{P \times L \times A \times N}{33{,}000}$$

where ihp = indicated horsepower

P = mean effective pressure of steam, psi, as found by a steam-engine indicator. In this case the mean effective pressure (average steam pressure throughout the stroke) will be 100 psi, since the inlet of steam is not cut off before the end of the stroke in most simple steam pumps

L = length of stroke, ft

A = area of steam piston, sq in.

N = number of working strokes per min

Applying this formula,

$$ihp = \frac{100 \times 1 \times 50.24 \times 50}{33{,}000} = 7.6 \ ihp \quad Ans.$$

Q If the pump in the previous question is delivering 330 gal of water per min to a height of 60 ft, what useful horsepower is being exerted?

A Useful hp $= \dfrac{330 \times 8.33 \times 60}{33{,}000} = 5.0$ hp   *Ans.*

Q What is the mechanical efficiency of the above pump?

A Mechanical efficiency $= \dfrac{\text{hp output}}{\text{hp input}} = \dfrac{5}{7.6}$

$$= 0.658 \text{ or } 65.8 \text{ percent} \quad Ans.$$

Q What is *piston speed?*
A *Piston speed* is the distance traveled by a pump piston in feet per minute. In a simplex pump the piston speed and the total piston travel per minute are the same. In a duplex pump, piston speed is the speed of one piston in feet per minute or one-half of the total piston travel per minute.

Q What are the piston speed and total piston travel in a duplex pump having an 8-in. stroke and making a total of 120 strokes per minute?
A Piston speed $= \dfrac{8 \times 60}{12} = 40$ ft per min   *Ans.*

Piston travel $= 40 \times 2 = 80$ ft per min   *Ans.*

Q What is the most suitable piston speed for a reciprocating pump?
A This depends entirely upon the size of the pump and the purpose for which it is used. Speeds should vary for different kinds of work and should be low enough, in the first place, to allow for wear and slip at this speed while supplying all wants. Then in an emergency, the capacity can be greatly increased by speeding up the pump. The following table gives the piston speeds recommended by a well-known pump manufacturer for various strokes and types of service.

Q What is the capacity in cubic inches of the water cylinder of a reciprocating pump if the cylinder diameter is 6 in. and the stroke is 8 in.?
A First find the area of the pump piston in square inches, using the common formula, area $= 0.785 \, D^2$. Then

$$\text{Area of piston} = 0.785 \times 6 \times 6 = 28.26 \text{ sq in.}$$
$$\text{Volume} = \text{area in sq in.} \times \text{length of stroke}$$
$$= 28.26 \times 8 = 226.08 \text{ cu in.} \textit{Ans.}$$

TABLE 18-1 Average Piston Speeds, in Feet per Minute, for Direct-acting Pumps

Stroke, in.	Boiler feed	General service	Low service
3	18	28	25
4	22	34	30
5	24	38	34
6	26	42	36
8	30	48	44
10	38	58	52
12	40	60	54
15	50	75	68
18	60	90	80

Q What is the capacity of this pump in gallons per minute if the pump makes 90 strokes per minute?
A If there is no slip,

$$\text{Capacity} = \frac{226.08 \times 90}{231} = 88 \text{ gal per min} \quad \textit{Ans.}$$

Q What is the actual capacity of this pump if the "slip," or leakage of water past the valves and pistons, amounts to 8 percent of the theoretical capacity?
A If the slip is 8 percent, then the actual capacity is $100 - 8 = 92$ percent of the theoretical capacity and

$$\text{Actual capacity} = \frac{88 \times 92}{100} = 80.96, \text{ say } 81 \text{ gal per min} \quad \textit{Ans.}$$

Q The discharge pipe from a pump at the bottom of a mine shaft, 460 ft deep, has an internal diameter of 4 in. What will be the static pressure on the pump piston in psi when the pipe is standing full of water, and what will be the total weight of water in the pipe?
A Static pressure on pump piston $= 460 \times 0.434$
$$= 199.64, \text{ say } 200 \text{ psi} \quad \textit{Ans.}$$
Area of pipe $= 4 \times 4 \times 0.785 = 12.56$ sq in.
Volume of water $= (12.56 \times 460) \div 144 = 40.12$ cu ft
Weight of water $= 40.12 \times 62.4 = 2,503$ lb   *Ans.*

Q How much water will be discharged in gallons per minute through a pipe of ½-in. internal diameter if the velocity of the water in the pipe is 4 ft per sec?
A 4 ft sec $= 4 \times 60 = 240$ ft per min
Area of discharge $= 0.5 \times 0.5 \times 0.785 = 0.196$ sq in.
Volume of discharge $= (0.196 \times 240) \div 144 = 0.326$ cu ft
Volume in gallons $= 0.326 \times 7.48 = 2.44$ gal per min   *Ans.*

Q How much water will be discharged in gallons per minute by a centrifugal pump with a 5-in. discharge and velocity at the discharge opening of 8 ft per sec?
A 8 ft per sec $= 8 \times 60 = 480$ ft per min
Area of discharge $= 5 \times 5 \times 0.785 = 19.6$ sq in.
Volume of discharge $= (19.6 \times 480) \div 144 = 65.3$ cu ft
Volume in gallons $= 65.3 \times 7.48 = 488$ gal per min   *Ans.*

Q How is the size of a reciprocating steam pump usually given?
A In giving the size we state the diameter of the steam cylinder, the diameter of the water cylinder, and the length of the stroke, all in inches and in the order mentioned. Thus, a pump having a 10-in.-diameter

steam cylinder, an 8-in.-diameter water cylinder, and a 12-in. stroke would be a $10 \times 8 \times 12$-in. pump.

Q An $8 \times 4\frac{1}{2} \times 10$-in. duplex pump makes a total of 80 strokes per minute. How many imperial gallons per minute does it discharge if slip is 10 percent?

A If the pump makes a total of 80 strokes per minute, each side makes 40 strokes per minute, and if slip is 10 percent of total capacity, the volumetric efficiency of the water end of the pump is 90 percent. Then

Water delivered in gal per min

= area of piston in sq ft × total piston travel in ft × gal in 1 cu ft × volumetric efficiency of pump

$$= \frac{4.5 \times 4.5 \times .785}{144} \times \frac{10 \times 80}{12} \times \frac{6.24}{1} \times \frac{90}{100}$$

$$= \frac{4.5 \times 4.5 \times .785 \times 10 \times 80 \times 6.24 \times 90}{144 \times 12 \times 100}$$

= 41.33 gal per min   *Ans.*

Q How would you calculate the size of a boiler feed pump for any given conditions?

A In the case of steam-driven boiler feed pumps of the reciprocating type, the steam end must be considerably larger than the water end to give a large reserve power capacity and enable the pump to work against high pressures. As boiler feed pumps usually handle hot water, it should be noted that hot water weighs less than cold water, and for pumping calculation purposes, its weight may be taken as, say, 60 lb per cu ft instead of 62.4. Piston speeds are also lower as a rule on boiler feed pumps than on general-service or low-service pumps.

The best procedure when calculating the size of a boiler feed pump for a given boiler installation is to assume a fairly low piston speed for normal operation, and then calculate the size of pump needed to deliver the required amount of water at this speed, making a generous allowance for wear and slip. If the capacity of the boilers is given in boiler horsepower, it should be remembered that 1 boiler hp is equivalent to the evaporation of 34.5 lb of water, from water at 212°F into steam at 212°F.

Q What will be the dimensions of an inside-packed double-acting duplex steam-driven pump to supply a battery of six 150-hp fire-tube boilers?

A In order to use the table of piston speeds, Table 18-1, we must assume a length of stroke. This is, of course, a matter of guesswork, but if we find at the end of our calculation that the stroke is not in proportion to the rest of the pump dimensions, we can easily change it to some more suitable length and check the problem over to find if any other changes in dimensions are necessary.

Assume that one side of the pump will supply half of the water and that the pump stroke will be 10 in., which gives us a suitable piston speed of 38 ft per min from the table. Also, allow 20 percent for wear and slip, which means that we are figuring on only 80 percent of the theoretical capacity being actually delivered by the pump. Then

Quantity of water to be supplied per hour by one side of the pump

$$= \frac{6 \times 150 \times 34.5 \times 100}{2 \times 80} = 19,406 \text{ lb}$$

Quantity of water required per minute $= 19,406 \div 60 = 323.4$ lb

Quantity of water required per stroke $= 323.43 \div \dfrac{38 \times 12}{10}$

$$= 7.093 \text{ lb}$$

Assume that the pump will be handling hot water weighing approximately 60 lb per cu ft. Then

Volume of water cylinder $= 7.093 \div 60 = 0.1182$ cu ft $= 204.2$ cu in.
Area of water cylinder $= 204.2 \div 10 = 20.42$ sq in.

Diameter of water cylinder $= \sqrt{\dfrac{20.4}{.785}} = \sqrt{26.0} = 5.1$ in., say $5\frac{1}{2}$ in.

Assume that the area of the steam cylinder is twice the area of the water cylinder, to give a good margin of power. Then

Diameter of steam cylinder $= \sqrt{\dfrac{2 \times 5.5 \times 5.5 \times 0.785}{0.785}}$

$$= \sqrt{60.5} = 7.78, \text{ say 8 in.}$$

Size of pump is $8 \times 5\frac{1}{2} \times 10$ in. *Ans.*

Q Calculate the size of simplex pump that would be needed under the same conditions as in the previous question.
A In this case we have only one cylinder, so it must be larger than the cylinders of the duplex pump and the stroke should also be longer. Assume a 12-in. stroke, which is given a piston speed of 40 ft per min in Table 18-1, and again allow 20 percent for slip. Then

Quantity of water required per hour $= \dfrac{6 \times 150 \times 34.5 \times 100}{80}$

$$= 38.812 \text{ lb}$$

Quantity of water required per minute $= 38,812 \div 60 = 647$ lb
Quantity of water required per stroke $= 647 \div 40 = 16.2$ lb.

Assume the pump handles hot water weighing 60 lb per cu ft. Then

Volume of water cylinder $= 16.2 \div 60 = 0.270$ cu ft $= 466$ cu in.

Area of water piston $= 466 \div 12 = 38.83$ sq in.

$$\text{Diameter of water piston} = \sqrt{\frac{38.83}{0.785}} = \sqrt{49.46} = 7.03, \text{ say 7 in.}$$

Assume the area of the steam piston is twice the area of the water piston. Then

$$\text{Diameter of steam piston} = \sqrt{\frac{2 \times 7 \times 7 \times 0.785}{0.785}} = 9.9, \text{ say 10 in.}$$

Size of pump is $10 \times 7 \times 12$ in. *Ans.*

Q What is meant by the *duty of a pump?*
A The *duty of a pump* is the foot-pounds of work done per 1,000,000 Btu supplied by the steam. It is used as a basis of comparison of pump performance.

Q During a pumping test, a duplex pump raised 18,400,000 lb of water against a head of 60 ft. The total heat supplied to the pump in the steam during the test was 30,600,000 Btu. What was the duty of the pump?
A Duty of pump $= \dfrac{18,400,000 \times 60}{30.6}$

$$= 36,078,000 \text{ ft-lb per } 1,000,000 \text{ Btu} \quad \textit{Ans.}$$

Q What would be the thermal efficiency of the pump in the previous question?
A Dividing foot-pounds of work done per million Btu by 778, converts foot-pounds to Btu, and this divided by 1,000,000 Btu gives the thermal efficiency. Thus,

$$\text{Thermal efficiency} = \frac{\text{work done per } 1,000,000 \text{ Btu} \div 778}{\text{heat equivalent of } 1,000,000 \text{ Btu}}$$

$$= \frac{36,078,000 \div 778}{1,000,000}$$

$$= \frac{36,078,000}{778,000,000} = 0.0464 \text{ or } 4.64 \text{ percent} \quad \textit{Ans.}$$

Q What is the difference between a "low" duty and a "high" duty pump?
A Large pumping engines which have compound cylinders and which use steam expansively have a fairly high thermal efficiency, say around 10 percent, and are classed as high-duty pumps. Small pumps that do not use steam expansively have a low thermal efficiency, around 2 to 3 percent, and are classed as low-duty pumps.

Q Calculate the horsepower that is expended by the water end of a reciprocating steam pump in forcing 450 gal of water per min to a height of 230 ft, assuming 10 percent of the power is used up in overcoming friction in the pump and pipe lines. What will be the required horsepower of the steam end if the mechanical efficiency of the pump is 85 percent?

A $\text{hp of water end} = \dfrac{\text{ft-lb of work done per min}}{33,000 \times 0.9}$

$= \dfrac{\text{lb of water per min} \times \text{height raised in ft}}{33,000 \times 0.9}$

$= \dfrac{450 \times 8.33 \times 230}{33,000 \times 0.9} = 29 \text{ hp} \quad Ans.$

$\text{hp of steam end} = \dfrac{29 \times 100}{85} = 34 \text{ hp} \quad Ans.$

Q The capacity of a pump is given as 4,860 U.S. gal per min; how much is this in imperial gallons?

A $\dfrac{4,860 \times 231}{277} = 4,053 \text{ imp. gal} \quad Ans.$

Q In the steam cylinder of a pump, the mean effective or average steam pressure is 60 psi, the length of the stroke is 14 in., the diameter of the cylinder is 12 in., and total number of strokes per minute is 80. Find the indicated horsepower.

A $ihp = \dfrac{PLAN}{33,000} = \dfrac{60 \times 14 \times 12^2 \times 0.785 \times 80}{12 \times 33,000} = 19.2 \text{ hp} \quad Ans.$

Q Find the diameter of a double-acting simplex pump to discharge 120 gal of water per min at a piston speed of 40 ft per min, allowing 10 percent extra for slip.

A $\text{Capacity} = 120 + 10 \text{ percent} = 132 \text{ gal per min}$

$\dfrac{132}{7.48} = 17.65 \text{ cu ft per min}$

$\text{Area of piston in sq in.} = \dfrac{\text{volume in cu ft}}{\text{piston speed in ft}} \times 144$

$= \dfrac{17.65}{40} \times 144 = 63.5 \text{ sq in.}$

$\text{Diameter of piston} = \sqrt{\dfrac{63.5}{0.785}} = \sqrt{81} = 9 \text{ in.} \quad Ans.$

Q How long would it take the pump in the previous question to empty a

sump 20 ft long, 10 ft wide, and 10 ft deep, standing full of water?
A Volume of water $= 20 \times 10 \times 10 = 2{,}000$ cu ft
$2{,}000 \times 7.48 = 14{,}960$ gal

$$\text{Time required} = \frac{14{,}960}{120} = 124.7 \text{ min, or } 2 \text{ hr } 4.7 \text{ min} \quad Ans.$$

Q How long would it take to empty this sump if a feeder of 50 gal per min is flowing into it?
A The amount by which water is actually lowered will be $120 - 50$ or 70 gal per min. Then

$$\text{Time required} = \frac{14{,}960}{70} = 213.7 \text{ min or } 3 \text{ hr } 33.7 \text{ min} \quad Ans.$$

Q Why are characteristic curves used for centrifugal pumps and what do they mean?
A Figure 18-3 shows characteristic curves of a centrifugal pump at constant speed. These curves are used because the relations between head, discharge, and pump speed are best determined by test rather than by calculation—unlike the same relations for positive displacement pumps (rotary and reciprocating). Horsepower and efficiency curves are included in Fig. 18-3 with the head-discharge curve for a pump running at constant speed. The head and capacity at which the pump is rated at this speed are those values obtained when the efficiency is a maximum. The value of the head at the point of maximum efficiency is usually less than at shutoff, but may be higher in exceptional cases.

The values used in plotting such curves are those computed for a given speed of the pump, usually that at which the pump operates at the best efficiency, or that at which it was designed to operate. For centrifugal

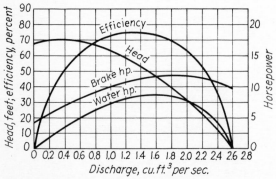

FIG. 18-3 Characteristic curves of a centrifugal pump at constant speed.

pumps, the relation between the various quantities is largely controlled by the angles and curvatures of the impeller blades and the shape of the volute, or the arrangement and design of the diffusion vanes. If the vanes are radial or inclined forward in the direction of rotation, the head will increase with increased capacity; if the vanes are curved backward sufficiently, the head will remain constant or decrease as the capacity increases, as shown in the curve.

In general, for a given centrifugal pump, the quantity of water delivered varies directly with the speed, the head with the square of the speed, and the power with the cube of the speed. Thus, doubling the speed of a pump impeller doubles the quantity of water pumped, which produces a head 4 times as great and requires 8 times as much power to drive the pump.

Because manufacturers furnish curves for their specific output, and as pump-performance curves vary within wide limits, just remember that general curves are for general approximations only. In Fig. 18-3, the highest point on the total head curve is 70 ft, thus the pump could not be used for a greater head, since at the existing constant speed no water would be delivered.

SUGGESTED READING

REPRINTS SOLD BY *Power* MAGAZINE
Pumps, 32 pp.
Power Handbook, 64 pp.

BOOKS
Hicks, Tyler, and Theodore Edwards: "Pump Application Engineering," McGraw-Hill Book Company, New York, 1970.
Elonka, Stephen M.: "Standard Plant Operators' Manual," 2d ed., McGraw-Hill Book Company, New York, 1975.
Elonka, Stephen M., and Orville H. Johnson: "Standard Industrial Hydraulic Questions and Answers," McGraw-Hill Book Company, New York, 1967.

19

PIPING, ACCESSORIES, AND CALCULATIONS

In the broad field of energy-systems engineering, pipes are arteries interconnecting machines. Without piping, today's energy systems—steam, compressed gas, air conditioning, refrigeration, liquid handling—could not exist. Piping and piping standards have been with us a long time. In 1820 the first standards appeared for cast-iron pipe in Britain; they were later adopted in America.

Here we cover piping details and calculations for stationary engineers who, although they do not design piping systems on a full-time basis, may find such information valuable for an occasional job. Before installing boiler-pipe connections, or any piping that will be subjected to severe stresses, always consult the ASME and other codes.

HOW PIPE IS MADE

Q What materials are used for making pipe?

A Pipe is made from clay, cement, concrete, lead, brass, copper, cast iron, wrought iron, wrought steel, alloys of different metals, and various plastic materials.

Q For what particular purpose is each kind of pipe used?

A Cement, concrete, and clay (tile) pipes are used mainly for water supply and drainage purposes. Lead pipe has a high resistance to corrosion and can be bent readily to any desired shape, but it cannot withstand high temperature or much internal pressure. Copper and brass pipes resist corrosion better than steel or iron but are more expensive and not so suitable for very high pressures. Copper pipe is very flexible and easily formed into coils or short-radius bends.

Cast-iron pipe is used largely for low-pressure and low-temperature purposes, such as water supply and drainage. It is weaker than wrought-iron or wrought-steel pipe, but less subject to corrosion. The addition of small amounts of nickel or chromium to cast iron increases its strength and resistance to corrosion considerably. Pure wrought-iron pipe resists corrosion better than wrought-steel pipe, and hence is preferred for purposes where this quality is important. However, most of the pipe used for high-pressure lines and general purposes is made from wrought steel. Where corrosive liquids have to be conveyed, steel pipes may be lined with lead or cement.

Pipe with high corrosion-resisting properties and able to withstand considerable internal pressure is made from a number of nonmetallic plastic compounds.

Q How are wrought-iron and wrought-steel pipes made?
A There are four common methods of making iron and steel pipe, namely: (1) butt welding; (2) lap welding; (3) drawing from the solid (seamless pipe); (4) straight or spiral riveting.

There are two *butt-welding* processes. In one, a thin strip of plate called a *skelp* is heated to the welding point in a furnace and then drawn directly from the furnace through a thimble or bell, as shown in Fig. 19-1. The edges of the skelp are beveled a little so that they meet squarely and weld together as they are brought into contact in the welding bell.

In the second butt-welding process, the skelp is not heated but formed cold into a tubular shape by passing it through a succession of rolls, the successive steps in bending the skelp being shown in Fig. 19-2. In the final set of rolls the edges are butted and fused together by electric or oxyacetylene welding. Figure 19-3 shows the resistance method of electric

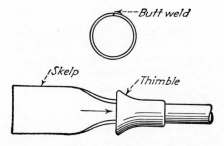

FIG. 19-1 Hot-process method of butt-welding pipe.

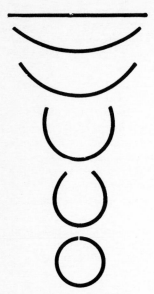

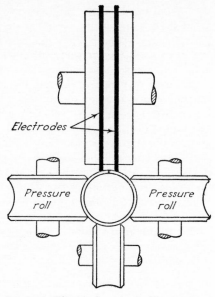

FIG. 19-2 Cold forming of skelp for butt-welding pipe.

FIG. 19-3 Electrical method of butt-welding pipe.

welding in which the metal at the joint is heated to the welding point by two electrodes as the edges are forced together by rolls. In the continuous cold-forming butt-welding process, the skelp is wound on a spool like a ribbon and the pipe is formed, welded, and cut off in lengths in one continuous operation.

In the *lap-welding* process, the edges of the skelp are "scarfed," Fig. 19-4, so that they overlap when the skelp is rolled into tubular form. The skelp is then heated to the welding point and drawn through rolls that press and weld the overlapping edges together.

FIG. 19-4 Form of skelp for lap-welding pipe.

Solid-drawn or *seamless* pipe is made by piercing a hole in the center of a red-hot billet of steel and then drawing the billet over a mandrel and through a succession of rolls that reduce the wall thickness until the finished size is reached, Fig. 19-5. The final finishing may be done while the tube is still hot or after it has cooled. Cold rolling makes the metal hard and

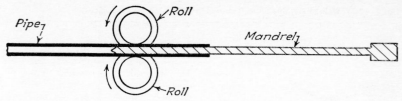

FIG. 19-5 Solid-drawing process of pipe manufacture.

brittle, and it must be annealed to restore its toughness and ductility.

Q What are the various grades of wrought-iron and wrought-steel pipe?
A Wrought-iron and wrought-steel pipe were formerly divided into three grades known as Standard, Extra Strong, and Double Extra Strong, but this classification has been superseded by one drawn up by the American National Standards Institute, formerly the American Standards Association. The ANSI specification consists of 10 different grades or schedules numbered as follows: 10, 20, 30, 40, 60, 80, 100, 120, 140, and 160. These grades are based on pressure-stress ratios which relate the pipe dimensions directly to the maximum pressure it is intended to carry. The outside diameter is the same for each size in all schedules, but the wall thickness varies, increasing gradually from the lightest grade, schedule 10, up to the heaviest grade, schedule 160. Pipe sizes go by the nominal inside diameter up to and including 12 in. Above 12 in. the outside diameter is taken, hence the term *O.D.* used for the larger sizes. ANSI dimensions for pipe sizes from ⅛ in. to 12 in. inclusive, in schedules 40 and 80, are given in Table 19-1.

The new grading makes a wider range available in the larger pipe sizes, but there is very little change in the dimensions of the smaller sizes. Up to and including 8 in., schedules 40 and 80 correspond to the old Standard and Extra Strong grades. There is no size corresponding to the old Double Extra Strong, and pipe in the heaviest grade, schedule 160, is still much lighter in weight and larger in bore than the Double Extra Strong.

Latest editions of standards may be purchased from American National Standards Institute, 70 East 45th St., New York, N.Y., 10017.

Q How would you determine which grade of pipe to use for any specific purpose?
A The proper grade to use will depend upon the service for which it is intended. Pipe in schedule 40 (old Standard grade) is most commonly used for general purposes but pipe in schedule 80 (old Extra Strong), or heavier if necessary, should be used in high-pressure work.

TABLE 19-1   Welded and Seamless Pipe Dimensions

Nominal size, in.	External	Internal Schedule 40	Internal Schedule 80	No. of threads per in.	Depth of thread, in.	Length of effective thread, in.	Size of tap drill for pipe taps
$1/8$	0.405	0.269	0.215	27	0.029	0.264	$21/64$
$1/4$	0.540	0.364	0.302	18	0.044	0.402	$29/64$
$3/8$	0.675	0.493	0.423	18	0.044	0.408	$19/32$
$1/2$	0.840	0.622	0.546	14	0.057	0.534	$23/32$
$3/4$	1.050	0.824	0.742	14	0.057	0.546	$15/16$
1	1.315	1.049	0.957	$11/2$	0.069	0.683	$1\,3/16$
$1\,1/4$	1.660	1.380	1.278	$11/2$	0.069	0.707	$1\,15/32$
$1\,1/2$	1.900	1.610	1.500	$11/2$	0.069	0.723	$1\,23/32$
2	2.375	2.067	1.939	$11/2$	0.069	0.756	$2\,3/16$
$2\,1/2$	2.875	2.469	2.323	8	0.100	1.137	$2\,11/16$
3	3.500	3.068	2.900	8	0.100	1.200	$3\,5/16$
$3\,1/2$	4.000	3.548	3.364	8	0.100	1.250	$3\,13/16$
4	4.500	4.026	3.826	8	0.100	1.300	$4\,3/16$
5	5.563	5.047	4.813	8	0.100	1.406	$5\,1/4$
6	6.625	6.065	5.761	8	0.100	1.512	$6\,5/16$
8	8.625	7.981	7.625	8	0.100	1.712	
10	10.750	10.020	9.564	8	0.100	1.925	
12	12.750	11.938	11.376	8	0.100	2.125	

PIPE THREADS

Q What is meant by *depth of thread, lead,* and *pitch,* in connection with pipe threads?

A *Depth of thread* is the depth to which the thread is cut, measured at right angles to the axis of the pipe. *Lead* is the distance that the screw advances in one turn. *Pitch* is the distance from center to center of adjacent threads. In a single-threaded screw, lead and pitch are the same. Pipe threads are all single-threaded. Double- and treble-threaded screws are used as transmission screws on machine tools and other machinery.

Q What form of thread is used on screwed pipe?

A The form of thread commonly used is known·as the American Standard taper pipe thread. It is V-shaped, with a 60° included angle, but the sharp point of the V is flattened at the top to a depth equal to 0.033 of the pitch, and flattened at the root of the thread to a similar amount. This makes the depth of the thread equal to 0.8 of the pitch.

The Whitworth form of thread is used in Great Britain and other countries. It is also a V-shaped thread with a 55° included angle, and with the points of the vee rounded off at the point and root of the thread. The depth of the Whitworth thread is 0.64 of the pitch.

The plain V thread is simpler than either of the foregoing, but it is difficult to keep sharp V points on taps and dies. Flattening or rounding the points of the threads increases the life of the threads and thread-cutting tools without weakening the thread to any appreciable extent.

Q Explain how pipe threads make tight connections when screwed together.

A Pipe threads are cut with a taper of ³/₄ in. per ft, or ¹/₁₆ in. per in., in order that they may make mechanically strong and gastight connections when the taper on the pipe tightens in the taper in the fitting. Because of the taper, a pipe thread is not of perfect form throughout its whole length. The nature of this imperfection is shown in Fig. 19-6, and the length of effective thread is given by the formula

$$L = \frac{1}{n} (0.8D + 6.8)$$

where D = actual external diameter of the pipe

n = number of threads per inch

The remaining threaded part has a few shallow threads that have little mechanical or jointing value.

Q What is the length of effective thread on a 1-in. pipe?

A From the formula,

$$\text{Length} = \frac{1}{11\frac{1}{2}} (0.8 \times 1.315 + 6.8) = \frac{2}{23}(1.052 + 6.8)$$

$$= \frac{2}{23} \times 7.852 = 0.68 \text{ in., or a little more than } \frac{5}{8} \text{ in.} \quad Ans.$$

Q What precautions should be observed when cutting and threading pipe?

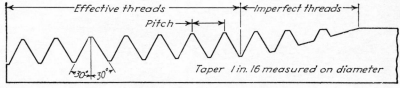

FIG. 19-6 Standard pipe-thread taper for self-sealing.

A Pipe should be cut with a hacksaw, parting tool, or power saw in preference to a wheel cutter, since the last tool leaves a burr on the inside of the pipe and some wheel cutters leave a burr on the outside as well. A burr on the outside makes threading more difficult unless it is ground or filed off. A burr on the inside reduces the effective area of the pipe unless it is reamed out. Quite a clean cut can be made by a cutting torch in the hands of a skilled operator.

When threading a pipe, care should be taken to cut the thread the proper length and depth. Threading machines and most hand dies are provided with guide marks and stops, and the threads are automatically cut to proper dimensions if the dies are in good shape and properly set. Blunt dies require much more power to turn than sharp dies, and cut a poor thread that will give trouble sooner or later. If threads are not cut deep enough or are too short, the pipe will enter only a few threads into the fitting, making a weak joint that will be almost sure to leak or break in a very short time. If threads are cut too deep or too long, the pipe will go too far into the fitting and tighten only on the last few threads. It may also block other openings into the fitting or butt against a valve seat and damage it.

After a pipe is threaded, it should be tapped sharply to remove cuttings from the threads, and a standard fitting should be tried on to make sure that the thread is properly cut. If a jointing paste is used, it should be put on the pipe threads and not in the fitting. Paste smeared in a fitting is simply pushed ahead of the pipe when it is screwed in, and washed into the system. Very little paste is required if the threads are good. Its main function is to lubricate the threads when the joint is being tightened and enable it to be broken when necessary, without stripping the threads. Keep dies well oiled and clean when cutting, and store them in a proper place to prevent damage to the cutting edges when they are not in use.

PIPE FITTINGS

Q What materials are used in the manufacture of pipe fittings?

A Pipe fittings are made from cast iron, alloy cast iron, bronze, malleable

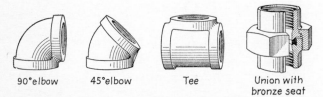

90°elbow   45°elbow   Tee   Union with bronze seat

FIG. 19-7 Standard threaded pipe fittings come in these angles and designs.

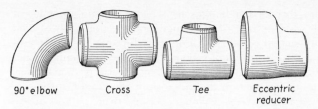

90°elbow   Cross   Tee   Eccentric reducer

FIG. 19-8 Pipe fittings for welded joints come in many shapes.

iron, cast steel, alloy cast steel, and forged steel. Chromium, nickel, and molybdenum are the usual alloying metals, and they are added to increase strength and resistance to corrosion. Fittings are made from all of these materials in a variety of weights to suit different purposes, but in general, cast iron and bronze are used for low and medium pressures and temperatures, alloy cast iron and malleable iron for medium pressures and temperatures, cast steel and alloy cast steel for fairly high pressures and temperatures, and forged steel for very high pressures and temperatures. Note that when choosing fittings for any particular purpose, the temperature to which the fittings will be exposed must be taken into account, as well as pressure.

Q Sketch some common forms of pipe fittings.
A Some of the commonly used types of screwed pipe fittings are shown in Fig. 19-7.

Q Sketch some common forms of welding pipe fittings.
A Some types of fittings for welded connections are shown in Fig. 19-8. These fittings are seamless, presenting a smooth inner and outer surface. The ends are beveled as shown for V welding.

Q Sketch and describe some common forms of flanged pipe joints.
A Sectional views of three typical flanged joints are shown in Fig. 19-9.
 In the *screwed flange* joint the pipe is threaded in the usual way, and the flanges are bored out and threaded with a taper so that they will be a

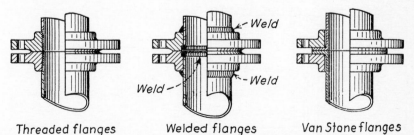

Threaded flanges   *Welded flanges*   *Van Stone flanges*

FIG. 19-9 Pipe flanges of the more popular designs.

tight fit on the pipe thread. The flanges are screwed on until the ends of the pipe are flush with the face of the flange or projecting slightly. The projecting ends are smoothed off by grinding or chipping and filing, and sometimes the ends are expanded a little in the flange by peening—that is, hammering the inside of the pipe lightly with a ball-peen hammer. Peening should not be necessary if the threads are good and fit properly.

The *slip-on welded flange* is a plain flange, slipped over the end of the pipe and welded inside and outside as shown.

The *Van Stone joint* is made by boring the flange out to slip easily over the end of the pipe, then flanging the end of the pipe and facing off the flanged end to provide a smooth surface for the gasket. Being loose, the flanges can be turned to any position for matching of bolt holes. In a modification of this joint the edges of the flanged ends of the pipes are seal-welded after the bolts are tightened, thus preventing any possibility of leakage. The screwed and slip-on welded flanges are suitable for low and moderate pressures, and the Van Stone and welding neck flanges (not shown) are preferable for very high pressures.

Q How are pipes connected by welding alone?
A The ends of the pipes are beveled to an angle of 30°, then butted together, and the V joint filled by a suitable welding rod and gas or electric welding apparatus. The use of welding rings, as shown in Fig. 19-10, keeps the ends of the pipes in proper alignment and strengthens the joint.

Q How are flanged joints made steam-, gas-, or water-tight?
A This is done by placing a gasket or ring between the joint faces and tightening the flange bolts, taking care to pull them up evenly all around so that there is an equal pressure on the ring or gasket at all points. Gaskets may be made from sheet rubber, rubber with wire insertion, sheet asbestos, corrugated soft-metal rings, or various other materials, depending upon the purpose of the pipe line and the pressure to be

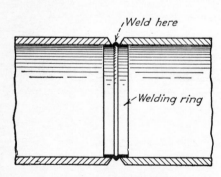

FIG. 19-10 Welding ring used in piping.

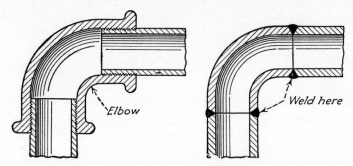

FIG. 19-11 Comparison of screwed and welded pipe joints.

carried. Metal rings of oval or octagonal cross section that fit into grooves in the faces of the flanges are often used on high-pressure lines, and sometimes flanges are made with a projecting ring on one flange which fits into a similar recess in the companion flange, a gasket being placed in the recess to make the joint tight. No gasket is needed when the flanges are seal-welded.

Q Has a welded joint any advantages over a screwed joint?

A When properly made, welded joints are more permanent than screwed joints and less likely to give trouble in course of time, but the making of welded joints requires more skill and judgment than the joining together of screwed fittings. Factory-welded jobs such as large pipe headers are probably more reliable than screwed joint headers, but for general erection and repair in the field, welding alone is not particularly safe unless it is done by thoroughly competent and experienced welders. People can be trained easily and quickly to make up screwed joints, but good welders are not trained in a day.

In general, welded joints have a number of advantages over screwed joints. Joints can be built up at any angle without the use of special fittings. Pipes do not have to be in exact alignment, which may save considerable time when laying large pipes. Connections can be made to existing lines without having to disconnect them. If the welding is properly done, there should be no projection on the inside of the pipe or fitting and therefore less resistance to the smooth flow of gas or liquid than that offered by the screwed type of fitting. This is shown clearly in Fig. 19-11. There is a saving of space occupied and metal used in places where a great number of joints have to be made, if welded joints can be used instead of screwed joints.

Q What other forms of pipe joints are used in addition to screwed and welded joints?

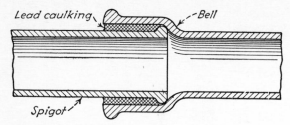

FIG. 19-12 Calked bell used in spigot joint.

A Four other common forms are (1) calked bell-and-spigot joint; (2) screwed bell-and-spigot joint; (3) Dresser joint; (4) soldered joint.

The *calked joint,* Fig. 19-12, is used mainly for drainpipe and low-pressure water supply. The pipes are of cast iron and the joint is made by inserting a spigot end into a bell end, calking a few rounds of oakum in the neck of the bell to stop the lead from running into the pipe, plastering clay or putty around the mouth of the bell, leaving a small opening on top, and pouring molten lead into the joint through this opening. After the lead has cooled, it is calked tight with a calking tool. The joint may also be made with lead wool instead of molten lead. The lead wool method is very useful where pipes have to be calked under water or with water running through them.

The *threaded bell-and-spigot joint* is also used on cast-iron pipe. The pipe joint is not calked in this case but screwed together, as shown in Fig. 19-13.

In the *Dresser joint,* Fig. 19-14, a packing ring assembly is used, consisting of two iron rings with a ring of soft packing between. This is slipped over the plain end of the pipe before it is inserted in the bell of the other pipe. Lugs in the outer iron ring pass through grooves in the flange of the bell, and a slight turn causes the lugs to enter a groove in the interior of the bell flange, thus locking the entire assembly in place.

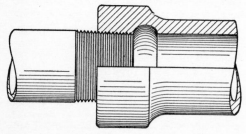

FIG. 19-13 Threaded bell in spigot joint.

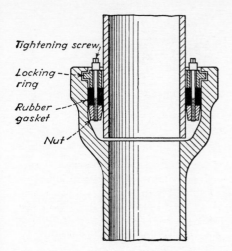

Tightening screw,

Locking--
ring

Rubber ---
gasket

Nut

FIG. 19-14 Dresser pipe joint.

Tightening the long cap screws, which pass through the packing ring and screw into the inner iron ring, brings the two iron rings closer together and squeezes the soft packing ring so that it presses hard against the spigot and bell, making a tight joint that permits considerable movement without leakage.

Soldered joints are used on copper pipe in plumbing work. The ends of the pipes and the inside of the fittings are tinned, and the joint is made by slipping the fitting over the end of the pipe and heating it with a blow torch to cause the solder to flow.

BENDS AND SUPPORTS

Q How are pipe bends made?

A Small-diameter pipe can often be bent cold by a roller-bending machine, or slight bends may be made by bending the pipe over some curved object, but large-diameter pipe must be heated before it can bend properly. If apparatus for heating and bending the pipe in one operation is not available, the usual practice is to heat a small portion of the pipe at a time and bend it a little at the heated part, continuing the process until the desired bend is secured. Care must be taken not to bend the pipe too much at one time when using this method, or the walls may be squeezed out of shape. Filling the pipe with dry sand and plugging the ends help to keep it in its proper shape while it is being bent, but all important large-size bends should be ordered from a pipe manufacturer or engineering works equipped to do this work.

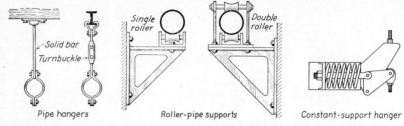

FIG. 19-15 Various methods of supporting piping for unrestricted movement as it expands and contracts.

Q How are pipe lines supported and anchored?

A Pipe lines are supported by steel or timber trestles, concrete pillars, cast-iron or wrought-steel wall brackets, or steel hangers of various designs. They are anchored—that is, held rigidly at one or more points—by clamps attached to supporting columns or brackets.

Figure 19-15 shows adjustable and nonadjustable hangers, single- and double-roller bracket supports, and a constant-support type. The first hanger shown has no adjustment for length but the second hanger is provided with a turnbuckle. This turnbuckle has right- and left-hand threads so that it can be lengthened or shortened for proper length adjustment and even distribution of weight.

The rollers on the bracket supports permit the pipe to move freely as it expands and contracts. The upper roller on the double-roller support holds the pipe down on the lower roller while still allowing free lengthwise movement.

In the last type of pipe hanger, the load is carried by a heavy coil spring. Means are provided for adjusting the compression of the spring, and an indicator on the side of the case enclosing the spring shows the load that is being supported by the hanger, in pounds, and the extension of the spring in inches.

EXPANSION

Q What provision must be made for taking care of expansion of steam lines?

A Expansion is taken care of by *expansion bends* and *joints*. The pipe line is anchored at certain points where excessive movement due to expansion is undesirable, and bends or expansion joints are placed in the line between these anchored points. As a rule, pipe lines are anchored close to points where they connect to boilers or engines, and the devices for the absorption of expansion are inserted in the long runs. Supports at points

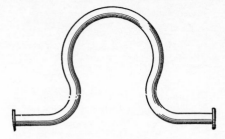

FIG. 19-16 Large expansion pipe bend.

other than anchors are designed to permit free movement of the pipes. If adequate provision is not made for expansion, there is danger of rupture of pipes, valves, or fittings, with disastrous consequences.

Q Sketch a good type of expansion pipe bend.
A A good form of expansion bend is shown in Fig. 19-16. If the bends are made to a large enough radius, this form of expansion pipe bend will take care of considerable expansion in a pipe line, without throwing an excessive stress on the flanges or other fittings

Q Sketch and describe several forms of expansion joints for pipe lines.
A The expansion joint shown in Fig. 19-17 consists of a smooth sleeve which slides in or out through a packing box as the pipe line expands or contracts in length. The packing box is packed with soft packing, held in place by a gland and studs. Tie rods guard against the possibility of the two halves of the joint pulling apart, and flanges are provided for connecting the joint into the pipe line.

The expansion joint shown in Fig. 19-18 also consists of a smooth sleeve sliding through a packing box, but it has internal traverse stops to guard against pulling apart instead of outside tie rods, and metallic packing instead of soft packing. It is more suitable for high pressures than the joint shown in Fig. 19-17.

Corrugated pipes and metal bellows are also used in the construction of expansion joints. An expansion joint with a welded metal bellows is shown in Fig. 19-19. Since the bellows is welded to the two halves of the joint, it is perfectly steamtight, and no packing box is necessary.

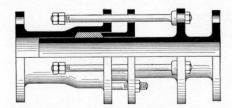

FIG. 19-17 Low-pressure expansion joint with restraining bolts.

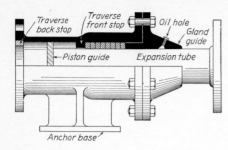

FIG. 19-18 High-pressure expansion joint with metallic packing.

Q How can the probable amount of expansion in a pipe line be calculated?
A The expansion can be calculated fairly closely for moderate temperature ranges if we know the coefficient of expansion of the metal from which the pipes are made. The coefficient of expansion is the amount of increase of length per unit of length, per degree Fahrenheit rise in temperature, or

> Expansion in in. = original length in in. × rise in temp. in °F
> × coefficient of expansion

Q What would be the amount of expansion in a steam pipe line, 340 ft long, if the outside temperature is 60°F and the pipe is carrying steam at 150 psi gage?
A A coefficient of expansion of 0.0000068 is given by various authorities for the metals from which steel and iron pipe are manufactured. Using this value, and taking temperature of steam as 366 F.,

$$\text{Expansion} = 340 \times 12 \times (366 - 60) \times 0.0000068$$
$$= 340 \times 12 \times 306 \times 0.0000068$$
$$= 8.49 \text{ or approximately } 8\frac{1}{2} \text{ in.} \quad Ans.$$

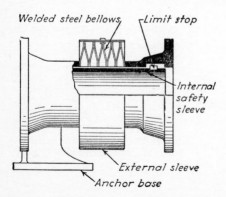

FIG. 19-19 Metal-bellows expansion joint needs no packing, does not leak.

STEAM SEPARATORS

Q What is a steam separator?

A A steam separator is an apparatus for separating out moisture that may be carried in suspension by steam flowing in pipe lines and for preventing this moisture from reaching and perhaps damaging engines, pumps, or other machinery that may be driven by the steam. The separation of the moisture from the steam is accomplished either by giving the steam a whirling motion or causing it to strike baffles that change the direction of its flow.

Q Sketch and describe two types of steam separator.

A A *centrifugal* steam separator is shown at the left in Fig. 19-20. In passing through this separator, the steam strikes a spiral-shaped plate which gives it a whirling motion. The moisture is thrown outward by centrifugal force, and it drains down the walls to the bottom of the separator while the steam passes up the central pipe and out to the line.

In the *baffle type* of steam separator shown at the right, the direction of flow of the steam is changed by suitable baffles. The moisture adheres to the baffle walls and drains down to the bottom of the separator. In all types of separators the water is drained away through a steam trap.

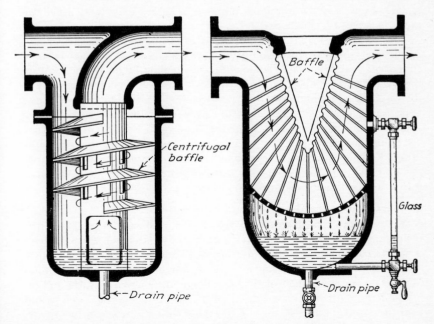

FIG. 19-20 Steam separators throw out droplets of condensate.

STEAM TRAPS

Q Name some of the common types of steam traps and explain how they operate.

A Traps may be roughly divided into six classes: bucket traps, float traps, tilting traps, expansion traps, thermostatic traps, and impulse traps.

In the *bucket* trap, the movement of an open bucket as it fills with condensate causes the discharge valve to open, and the steam pressure then forces the condensate out of the trap. As the bucket empties, it returns to its original position and closes the discharge valve.

Float traps contain a float, usually a hollow metal ball. As the trap fills with condensate, the float rises and opens the discharge valve. As the steam pressure empties the trap, the float falls and closes the discharge valve.

In the *tilting* trap the condensate drains into a counterbalanced cylindrical- or spherical-shaped vessel, and the weight of water causes the vessel to tilt and in doing so to open a discharge valve. When the steam pressure has forced out the water, the counterbalance brings the tilting bowl back to the closed position.

Expansion traps may be actuated by the expansion and contraction of some metal element in the trap, by oil in a sealed cartridge acting upon a sealed plunger, or by a volatile liquid in a sealed container, usually a metal bellows to facilitate expansion.

The *thermostatic* trap is a form of expansion trap. The discharge valve is opened or closed by the contraction and expansion of a metal bellows containing a highly volatile liquid with a low boiling point. When the trap is cold, the valve is open and water discharges freely. When steam enters the trap, the liquid in the bellows vaporizes. This causes the bellows to expand, since the vapor exerts a greater pressure than the liquid, and the expansion of the bellows closes the discharge valve.

The bucket trap is perhaps the most widely used trap for general purposes. Expansion traps, thermostatic traps, and combination float and thermostatic traps are extensively used in heating systems. The impulse trap is used on high-pressure steam lines.

Q What are *return* and *nonreturn* traps?

A *Return* traps are designed to return water of condensation to the boiler, against the boiler pressure. Tilting traps and some forms of float traps are commonly used for this purpose. They are set a few feet above the boiler and a live steam line is run from the boiler to the trap. Tilting of the bowl or movement of the float opens the discharge valve and at the same time opens a valve admitting live steam to the trap. The steam pressure from the live steam line, together with the few feet of gravity head from the trap to the boiler, causes the condensate to flow from the trap into the boiler.

The term *nonreturn* applies to traps of any type that have no live-steam connection direct from the boiler and that simply discharge the condensate to waste or to a hot-water receiver against little or no back pressure.

Q Sketch and describe two common types of bucket steam traps and explain how they operate.

A A common type of open-bucket trap is shown in section in Fig. 19-21. When water enters the trap, the bucket floats and presses upward on the discharge valve, keeping it closed so that no steam escapes. When the water rises above the level of the rim of the bucket, it fills the bucket, causing it to sink and open the discharge valve. The steam pressure acting upon the surface of the water in the bucket forces it up through the discharge valve. As the bucket empties, it again floats upward and closes the discharge valve.

The trap shown in Fig. 19-22 is also of the bucket type but with the bucket inverted. When the trap is filled with water, the bucket sinks to the bottom and opens the discharge valve, allowing the water to escape freely, but as soon as steam begins to enter the trap, the bucket fills with steam and air that may be carried with the steam, and floats upward, closing the discharge valve. As fresh condensate enters the trap, the steam and air are forced out of the small vent hole in the top of the bucket by the rising water level; the bucket loses its buoyancy and sinks to the bottom again, opening the discharge valve.

Q Sketch and describe some form of thermostatic trap.

A A form of thermostatic trap that is in very common use on steam-heating systems is shown in Fig. 19-23. The discharge valve of this trap is attached to a thin metal bellows or other thin-walled metal device containing a highly volatile liquid that will evaporate into a gas at a compara-

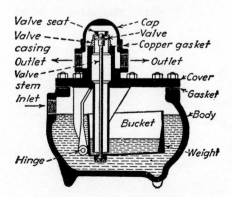

FIG. 19-21 Open-bucket steam trap.

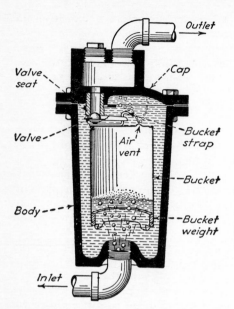

FIG. 19-22 Inverted-bucket steam trap.

tively low temperature. When the trap fills with condensate, the bellows or diaphragm contracts and opens the discharge valve so that the water may flow freely out of the trap, but when steam enters the trap, the volatile liquid in the bellows or other sealed container vaporizes, and the resulting expansion of the bellows or container diaphragm closes the discharge valve and prevents the escape of steam.

Q Explain the impulse steam trap.

A The impulse steam trap, Fig. 19-24, operates on the principle that hot water under pressure tends to flash into vapor when the pressure is reduced. When hot water is flowing to the trap the pressure in the control chamber is reduced, causing the valve to rise from its seat and discharge water. As steam enters the trap, the pressure in the chamber

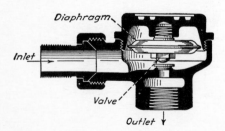

FIG. 19-23 Thermostatic trap.

increases, causing the valve to close, thus reducing the flow of steam to that which can pass through the small center orifice in the valve. It requires 3 to 5 percent of full condensate capacity to prevent steam from being discharged through this small orifice. Impulse traps are available for all pressures, they are compact, and the few parts subject to wear can be easily replaced. It is necessary to install a strainer, since small particles of foreign matter will interfere with the operation of the valve.

PIPING LAYOUTS

Q How are piping layouts usually represented in drawings, and how are measurements indicated on the drawings?

A There are two methods of representing pipe lines and fittings in drawings, *single-line* drawing and *double-line* drawing. The first method is commonly used in rough sketches, and the second is used in finished drawings made to scale.

Pipe measurements are taken from center to center of pipes and fittings, and the radius of a pipe bend is always measured to the center of the pipe. Examples of dimensioning are shown in Figs. 19-25 and 19-26.

Q Show how dimensions are inserted in double-line pipe drawings.

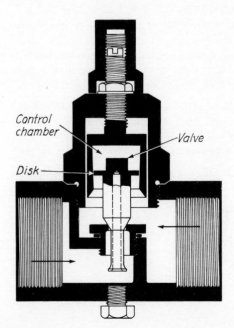

Control chamber

Valve

Disk

FIG. 19-24 Thermodynamic (impulse) steam trap.

A Figures 19-25 and 19-26 are reproductions of scale drawings, reduced from the original size. Figure 19-25 shows a short run of straight pipe with a valve and two elbows, and Fig. 19-26 shows a gooseneck bend.

Q Show some of the symbols used to represent valves and fittings in single-line pipe sketches, and give an example of a single-line pipe sketch.

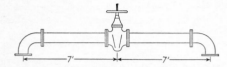

FIG. 19-25 Double-line drawing of pipe with valve and fittings.

A Figure 19-27 shows a few of the commonly used single-line symbols, and Fig. 19-28 shows their application.

Q Draw the piping layout for a battery of two dry-back fire-tube boilers.
A Fig. 19-29 shows the header supported by wall brackets and anchored midway between the boiler connections. Provision is also made for expansion by curving the main steam line and a steam trap is installed under the header for drainage.

CALCULATIONS

Q What are some common cases where exact lengths of pipe have to be calculated?
A Probably the most common cases requiring some arithmetical calculation are (1) finding length of a 45° offset; (2) finding length of a pipe bend; (3) finding length of spiral coil; (4) finding length of a helical coil.

Q How would you find the length of pipe required to make a 45° offset?
A This is calculated from the ratios between the lengths of the sides of a right-angled triangle. As shown graphically in Fig. 19-30, these ratios may be stated in two ways, $1:1:1.414$ and $0.707:0.707:1$.

Q What is the length of the 45° offset shown in Fig. 19-31, measured from center to center of fittings?
A Length of offset $= 16 \times 1.414 = 22.62$, say $22\frac{5}{8}$ in. *Ans.*

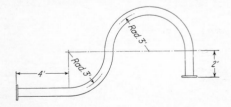

FIG. 19-26 Double-line drawing of pipe bend with flanges.

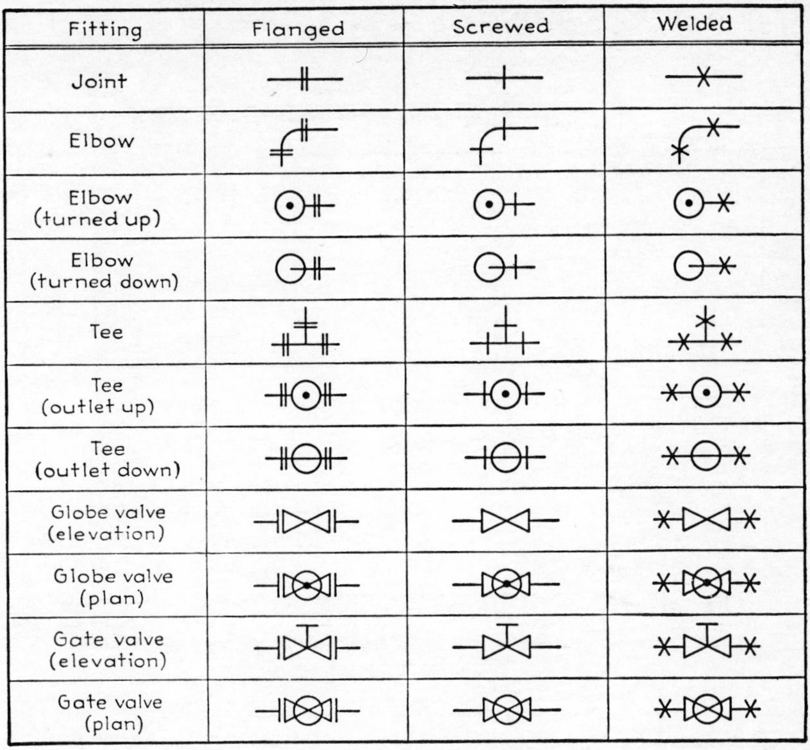

Fitting	Flanged	Screwed	Welded
Joint			
Elbow			
Elbow (turned up)			
Elbow (turned down)			
Tee			
Tee (outlet up)			
Tee (outlet down)			
Globe valve (elevation)			
Globe valve (plan)			
Gate valve (elevation)			
Gate valve (plan)			

FIG. 19-27 Standard single-line symbols for pipe fittings and valves.

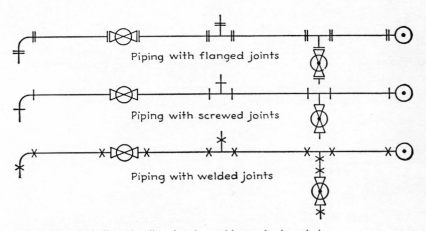

Piping with flanged joints

Piping with screwed joints

Piping with welded joints

FIG. 19-28 Single-line pipe-line drawings with standard symbols.

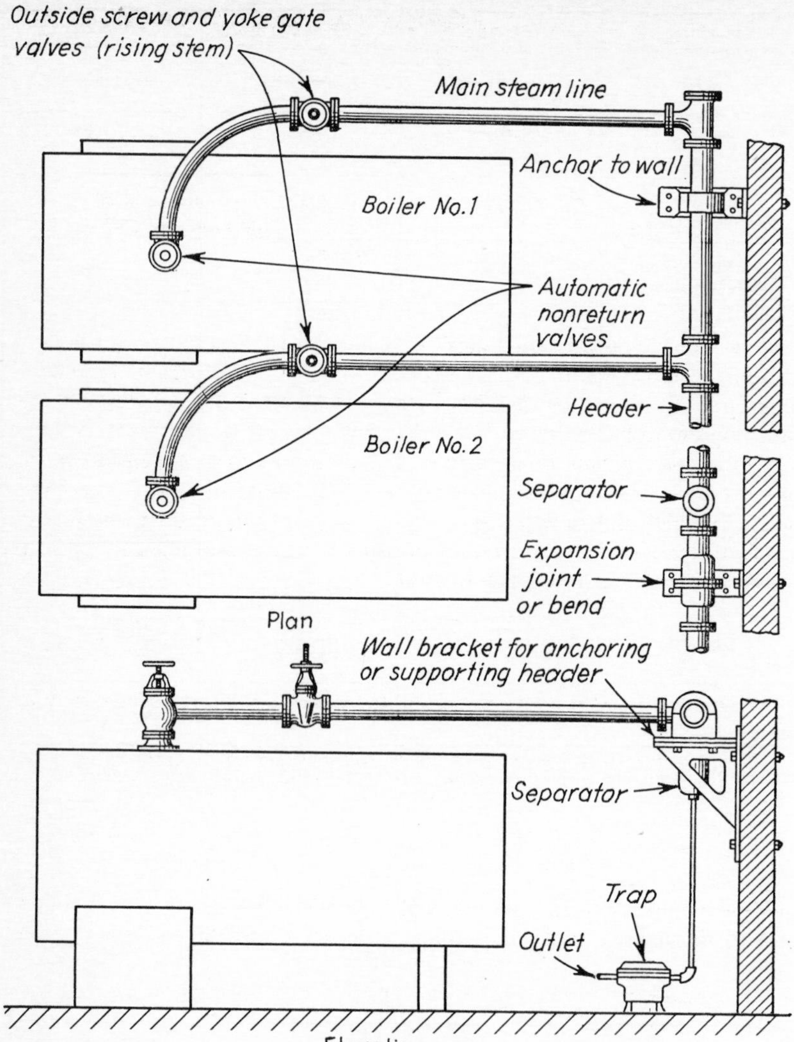

Outside screw and yoke gate valves (rising stem)

Main steam line

Anchor to wall

Boiler No.1

Automatic nonreturn valves

Header

Boiler No.2

Separator

Expansion joint or bend

Plan

Wall bracket for anchoring or supporting header

Separator

Trap

Outlet

Elevation

FIG. 19-29 Piping layout for battery of two boilers, plan and elevation.

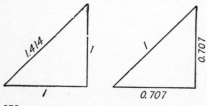

1.414

1

1

1

0.707

0.707

FIG. 19-30 Forty-five-degree right-angled triangle ratios for calculations.

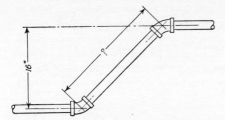

FIG. 19-31 Forty-five-degree offset in pipe line with 16-in. offset.

Q How would you find the length of a pipe bend?
A Bends are usually made in the form of an arc of a circle. A circle contains 360°. If its radius is known, the complete circumference can be found by multiplying the radius by $2\pi = 2 \times 3.14$.

In order to calculate the length of pipe required to make a pipe bend, we must know the angle of the sector of which the bend is an arc and the radius to which the pipe is to be bent. Then, by simple proportion, the length of the bend is to the circumference of the circle as the angle of the sector is to 360, or, stated as a formula,

$$\text{Length of bend} = \frac{\text{circumference of circle} \times \text{angle of sector}}{360}$$

Q What length of pipe is required to make the bend shown in Fig. 19-32?
A $\text{Length} = \dfrac{\text{circumference of circle} \times \text{angle of sector}}{360} = \dfrac{2\pi R \times 60}{360}$

$$= \frac{2 \times 3.14 \times 42 \times 60}{360} = 44 \text{ in.} \quad Ans.$$

Q How would you find the length of a spiral coil?
A Add the inner and outer diameters together and divide by two. This

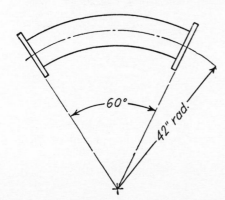

FIG. 19-32 Sixty-degree pipe bend.

gives the mean or average diameter. Multiply the mean diameter by π to get the average circumference, and then multiply the average circumference by the number of coils to get the total length of pipe in the coil.

This rule can be used to find the lengths of rolls of material such as belting and sheet lead, without having to unroll and actually measure the length. Diameters are measured from center to center of coils or rolls.

Q What length of 1-in. pipe will be required to make a spiral containing 10 coils, outside diameter of 36 in. and inside diameter of 8 in.?

A Mean diameter $= \dfrac{36 + 8}{2} = 22$ in.

Mean circumference $= 22 \times 3.14$
Total length $= 22 \times 3.14 \times 10 = 691$ in. $= 57$ ft 7 in.   *Ans.*

Q How would you find the length of a helical pipe coil?
A If the coils are close together, each coil is very nearly a true circle, and the error will be very slight if we assume the length of one coil in this case to be π (3.14) times the mean diameter. If the coils are not close together, the length of one coil will be a little more than $3.14 \times D$, and we must use the right-angled triangle rule as well as the formula for circumference in order to get the true length. The right-angled triangle rule is: "The square on the hypotenuse (longest side) of a right-angled triangle is equal to the sum of the squares on the other two sides."

First find the mean circumference by multiplying mean diameter by π. Then measure the *pitch* of the spiral, that is, the distance from center to center of adjacent coils. The pitch is one leg of the right angle and the mean circumference is the other, and

$$\text{Length of one coil}^2 = \text{mean circumference}^2 + \text{pitch}^2$$

or

$$\text{Length of one coil} = \sqrt{\text{mean circumference}^2 + \text{pitch}^2}$$

Q Find the length of one turn in the helical pipe coil shown in Fig. 19-33.
A Mean circumference of coil $=$ mean diameter $\times \pi$
$$= 12 \times 3.14 = 37.7 \text{ in.}$$

$$\text{Length of one coil} = \sqrt{\text{mean circumference}^2 + \text{pitch}^2}$$
$$= \sqrt{37.7^2 + 4^2} = \sqrt{1,437}$$

$$= 37.9 \text{ in., say 38 in.}   Ans.$$

Q How can we find the size of a steam pipe, given the weight of steam to be carried per hour in pounds, the pressure in pounds per square inch, and the velocity of flow in feet per minute?
A If the length of the pipe is so short that pressure drop may be

neglected, the required pipe area in square inches may be found by multiplying the volume of steam flowing in cubic feet per minute by 144 and dividing this by the velocity of steam flow in feet per minute. The pipe diameter can then be found by the usual formula for finding diameter from area. The volume of 1 lb of steam at the given pressure is found in the steam tables, Tables 16-3 and 16-4. Stated as a formula,

$$\text{Area of pipe in sq in.} = \frac{\text{volume of steam flowing in cu ft per min} \times 144}{\text{velocity of steam flow in ft per min}}$$

$$\text{Diameter of pipe in in.} = \sqrt{\frac{\text{volume of steam in cu ft per min} \times 144}{\text{volume of flow in ft per min} \times 0.785}}$$

Q What size of steam pipe would be required to handle a flow of 16,000 lb of steam per hr, at a pressure of 250 psia and a velocity of 8,000 ft per min?

A $\text{Area of pipe} = \dfrac{16,000 \times \text{cu ft per lb} \times 144}{60 \times 8,000}$

$\text{Diameter of pipe} = \sqrt{\dfrac{16,000 \times 1.84 \times 144}{60 \times 8,000 \times 0.785}} = \sqrt{11.25}$

$\qquad = 3.36$ in., which is the internal diameter of a schedule 80, $3\frac{1}{2}$-in. pipe. *Ans.*

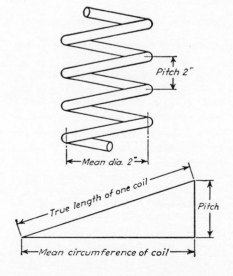

FIG. 19-33 Finding length of helical pipe coil.

Q How are sizes of pump suction and discharge pipes calculated?
A Sizes of pipes will depend upon velocity of flow; 500 ft per min for the discharge and 200 ft per min for the suction are fairly common values. Then, if

D = diameter of pipe, in.
A = cross-sectional area of pipe, sq ft
G = volume of water pumped per min, gal
V = velocity of flow, ft per min

$$G = A \times V \times 7.48 = \left(\frac{D}{12}\right)^2 \times 0.785 \times V \times 7.48$$

$$= \frac{D^2 \times 0.785 \times V \times 7.48}{144}$$

and

$$D = \sqrt{\frac{G \times 144}{0.785 \times V \times 7.48}} = 4.95 \times \sqrt{\frac{G}{V}}$$

Q What should be the sizes of the suction and discharge pipes of a direct-acting steam pump delivering 80 gal per min?

A Diameter of suction $= 4.95 \sqrt{\dfrac{80}{200}} = 4.95 \times 0.63 = 3.12$ in. *Ans.*

(A schedule 40, 3-in. pipe will do.)

Diameter of discharge $= 4.95 \sqrt{\dfrac{80}{500}} = 4.95 \times 0.4 = 1.98$ in. *Ans.*

(A schedule 40, 2-in. pipe will do.)

Q What general precautions should be observed in the erection of pipe lines?
A The proper size and grade of pipe, valves, and fittings should be selected for the service for which the line is intended. Joints, whether welded, screwed, or flanged, should be carefully made, with pipes in exact alignment. Threads on screwed pipe should be clean and smooth, and a good jointing compound should be used to preserve the threads, lubricate the joint while it is being tightened, and facilitate the breaking of the joint if it ever has to be taken apart. Faces of flanges should be clean and smooth, and bolt holes should be in line. Gasket material should be suitable for the pipe-line service. Sheet rubber, for example, may be excellent for a low-pressure water line, but it would be entirely unsuitable for a high-pressure steam line. Flange bolts should be tightened up evenly all around so as to keep flange faces exactly parallel. Pipe lines should be well supported, with an even distribution of weight on the

hangers or other supports, and anchored at suitable points and well drained by steam traps. There should be no low points where water can collect, but if this is unavoidable, traps should be placed at these places. Steam lines should be covered with insulating material to prevent loss of heat by radiation. The welding of pipe joints should be done only by experienced welders, so always check the ASME Welding Code.

Q Why are schemes for pipe-line identification desirable, and what methods of identification are used?

A If pipe lines are clearly marked in some manner that indicates the kind of material being conveyed through the line, they can be easily traced by persons not acquainted with the piping layout, and valves that will shut off the flow through any accidental leak or break can be quickly located and closed. Newcomers to the operating staff can also get the run of the plant in less time and be less likely to make costly mistakes if some system of marking pipe lines is used.

In some identification systems, the entire line is painted a distinctive color, different colors being used to represent steam, water, gas, etc. In other schemes wide color bands are painted at intervals. The colors used usually have some relation to the nature or use of the substance being carried through the line; thus, red is invariably chosen for fire-protection lines. Orange and yellow are common choices for lines carrying dangerous or poisonous materials, as they are easily distinguished even in a poor light. For safe materials, green or other less conspicuous colors may be used. Where pipe-identification schemes are in use, explanatory charts should be posted in places where they may be readily consulted in case of emergency.

THERMAL INSULATION (LAGGING)

Q Why should bare surfaces be insulated?

A By retarding the flow of heat, thermal insulation will: (1) reduce heat gain or loss from piping, equipment, and structures, thus decreasing the amount of heating or refrigeration capacity needed, (2) control surface temperature for personnel protection and comfort, (3) prevent icing or water vapor condensation on cold surfaces, (4) make it easier to control space and process temperature. Insulation can also (1) add structural strength to floors, walls, or ceilings, (2) prevent or retard the spread of fire; (3) reduce noise level; (4) absorb vibration. It is also a good support for surface finish and helps keep out water vapor if the joints are properly sealed.

Figure 19-34 shows a cross section of an insulated metal surface, which might be a steam drum, steam piping, or any hot surface. Temperature

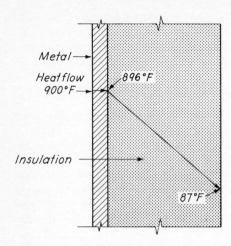

FIG. 19-34 Temperature drop through metal and insulation.

drop through metal is negligible, thus heat loss through bare metal is very great until it strikes the insulation, where it drops sharply.

Q What are some insulating materials used for high temperatures?

A Asbestos, alumina silica ceramic fiber, fibrous potassium titanate, diatomaceous silica, mineral fiber, expanded silica (perlite), felted glass fiber, glass (cellular), and 85 percent magnesia. The maximum temperature for alumina silica ceramic fiber is 2,300°F, and for 85 percent magnesia is 600°F. The others fall in between these limits.

SUGGESTED READING

REPRINTS SOLD BY *Power* MAGAZINE
Corrosion, 36 pp.
Mechanical Packing, 24 pp.
Piping, 16 pp.
Power Handbook, 64 pp.
Pumps, 32 pp.
Thermal Insulation, 24 pp.
Valves, 16 pp.

BOOKS
Elonka, Stephen M.: "Standard Plant Operators' Manual," 2d ed., McGraw-Hill Book Company, New York, 1974.

20

BOILER ROOM MANAGEMENT

Firemen or operating engineers are not worth their salt if they are not good housekeepers. That means maintaining equipment in proper repair and keeping log sheets and other records. Some machinery insurance companies furnish log sheets for boilers, refrigeration equipment, etc. Figure 20-1 is a log sheet for low-pressure steam heating boilers, requiring weekly readings by the operator.

RECORDS

Q In addition to the daily or weekly logs kept for a boiler room, what monthly checks must be made and recorded?
A Check safety valve by pulling try lever to open position with steam at full pressure. Release to allow valve to snap closed.

Check low-water fuel cutoff at least monthly by actually lowering the water level in the boiler slowly to simulate a developing low-water condition.

> CAUTION: Do *not* lower the water below bottom of water gage glass. Should cutoff not function properly under this test, it must be immediately overhauled and placed in operating condition.

Dismantle low-water fuel cutoff for complete overhaul at regular intervals. All the internal and external mechanism (including linkage, con-

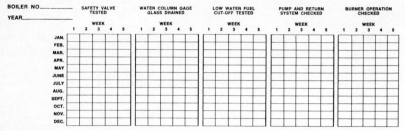

FIG. 20-1 Log sheet for low-pressure steam heating boiler requiring weekly readings.

tacts, mercury, bulbs, floats, and wiring) should be carefully checked for defects. Always refer to manufacturer's recommendations. Check boiler and entire system carefully and completely for leakage or other defects at regular intervals at pipe connections, flanges, traps, and valves. In the interest of good housekeeping, any unsatisfactory condition should be noted and corrected.

Service oil or gas burner controls—and thoroughly check all operating and protective burner controls—at least monthly. In general, this service should be obtained from a reliable outside service organization. If serviced by the operator, a complete record of work done should be entered in the log. First, check manufacturer's instructions and recommendations.

Clean fire side of boiler at least monthly. At the time of this cleaning, all brick work and refractory should be checked and repaired as needed.

Wash out boiler. The internal surfaces of the boiler should be checked at regular intervals to determine if scale or corrosion is present. Contact your boiler inspector if any unusual condition is noted. Length of time between washouts should be varied in accordance with conditions noted at time of washout.

Q What other records should be kept?
A The water analysis report, Fig. 20-2, is an example. Whether tests are made by operator or service organization, water analysis reports must be kept along with those of major repairs, such as new tubes rolled into a boiler, new brickwork, new boiler fittings installed, etc.

Records of inventory should also be kept on such items as gage glasses, boiler tubes, oil burner plates, packings, and gaskets (manhole, handhole plate) so that the operator isn't "caught short" when an emergency repair must be made. Waiting for critical components to be delivered causes costly downtime.

A repair record book should be kept to show *all* major repairs in the life of a boiler, recording each repair, by whom made, and the date. Such a report "history book" is as important as your medical history is to your doctor.

The operator in charge of the boiler room should keep a file drawer containing (*a*) blueprints of all major equipment in the boiler room, (*b*) manufacturer's literature of all equipment, with instruction sheets for safe operation, (*c*) fuel deliveries, (*d*) filled-in charts from recording instruments, etc.

Q Design a practical shift schedule for four firemen in a typical large boiler heating-plant.

A Figure 20-3 is for a plant of four firemen that has a chief and an assistant. If one fireman is sick or off work, then either the chief or his assistant must take the shift. The schedule gives each man a complete cycle of shifts and hours in a four-week period (underlines on chart). The man who works the 12-to-8 A.M. shift on Sunday has the sixth shift for that week, which means overtime for him.

This arrangement is more popular than a schedule of 8 hours on, 24

WATER ANALYSIS REPORT		Collected 4/26/75	
Sample No. 605		Analyzed 4/27/75	
For ABC Co.		Reported 5/4/75	
Ion		**epm**	**ppm as CaCO₃**
Cations Calcium as Ca	62 ppm	3.10	155
Magnesium as Mg	31 ppm	2.54	127
Sodium and potassium as Na	38 ppm	1.64	83
Total cations		7.28	365
Anions Bicarbonate as HCO₃	250 ppm	4.10	205
Carbonate as CO₃	0 ppm	0	0
Hydroxide as OH	0 ppm	0	0
Chloride as Cl	11 ppm	0.31	15
Sulfate as SO₄	138 ppm	2.87	145
Nitrate as NO₃	ppm		
Total anions		7.28	365

Silica as SiO₂	5 ppm	Total hardness	282 ppm CaCO₃
Iron as Fe₂O₃	1.2 ppm	Methyl orange alkalinity	205 ppm CaCO₃
Total dissolved solids	536 ppm		
Suspended solids (weight)	5 ppm	Phenolphthalein alkalinity	0 ppm CaCO₃
Chloroform-extractable matter	ppm		
Turbidity (after shaking)	5 ppm	pH 7.7 Color	
Carbon dioxide as CO₂	10 ppm	Sp conductance	μmhos

FIG. 20-2 Water analysis report is important boiler room record.

	Day of week		
Shift	**S M T W T F S**	**S M T W T F S**	**S M T W T F S**
First	B B B B C C C	C C C C A A A	A A A A D D D
Second	C C A A A A A	A A D D D D D	D D B B B B B
Third	D D D D D B B	B B B B B C C	C C C C C A A
First	D D D D B B B	B B B B C C C	C C C C A A A
Second	B B C C C C C	C C A A A A A	A A D D D D D
Third	A A A A A D D	D D D D D B B	B B B B B C C
First	A A A A D D D	D D D D B B B	B B B B C C C
Second	D D B B B B B	B B C C C C C	C C A A A A A
Third	C C C C C A A	A A A A A D D	D D D D D B B

FIG. 20-3   Shift schedule for four firemen during 4-week period.

hours off. First shift is from midnight to 8 A.M., second shift from 8 A.M. to 4 P.M., and the third shift from 4 P.M. to midnight. (For 12 additional shift schedules, see Chap. 18 in "Standard Boiler Operators' Questions and Answers.")

Q What entries should be made in a daily log book?
A At the beginning of each shift the boiler operator enters (a) blew down water column, (b) blew down gage glass, (c) operated try cocks, (d) lifted safety valve, except for boilers in the 1000-psi range, (e) blew down boilers, unless on a specially controlled schedule, and (f) examined boilers generally for leaks or unusual conditions. If some unusual condition arises, or if an accident occurs, the first thing a boiler inspector wants to see, if available, is a record of routine observations of boiler operators.

Q How should records of repairs be kept?
A Some operators keep records of repairs in full detail. For example, on a line drawing of the front of a typical sinuous-header water-tube boiler, all the tubes may be identified by giving each a number. Then a card record can be made of anything that happens to each one. This data becomes very valuable when a leak occurs at the rolled joint of a tube end. It won't take long to decide whether to attempt rerolling the tube or to go ahead with its replacement. The card records will show if the tubes were installed 15 or 20 years ago. If they were, it is very likely they will not stand much rerolling.

A card file is a valuable system of recording performance and maintenance of a boiler, pump, feed-water heater, or other machinery. The cards should be filed in alphabetical order in a box of convenient size or in a metal filing cabinet. Typical equipment cards should carry all notations of value in a well-designed maintenance program.

SUGGESTED READING

REPRINT SOLD BY *Power* MAGAZINE
Power Handbook, 64 pp.

BOOK
Elonka, Stephen M.: "Standard Plant Operators' Manual," 2d ed., McGraw-Hill Book Company, New York, 1975.

NATIONAL AUTHORITIES ON BOILERS

American Society of Mechanical Engineers, 345 E. 47th St., New York, NY 10017

American Boiler Manufacturers Association, 1500 Wilson Blvd., Arlington, VA 22209

Edison Electric Institute, 750 Third Ave., New York, NY 10017

National Board of Boiler and Pressure Vessel Inspectors, 1055 Crupper Ave., Columbus, OH 43229

National Fire Protection Association, 60 Batterymarch St., Boston, MA 02100

Uniform Boiler and Pressure Vessel Laws Society, 57 Pratt St., Hartford, CT 06103

APPENDIX

MEASUREMENTS _____

LENGTH

12 inches	=	1 foot
3 feet	=	1 yard
5,280 feet	=	1 statute mile
6,080 feet	=	1 nautical mile
1 mil	=	0.001 inch

AREA

144 square inches	=	1 square foot
9 square feet	=	1 square yard
Cross-sectional area in circular mils	=	Square of diameter in mils

VOLUME

1,728 cubic inches	=	1 cubic foot
27 cubic feet	=	1 cubic yard

WEIGHT

16 ounces	=	1 pound
2,000 pounds	=	1 short ton
2,240 pounds	=	1 long ton

CIRCULAR MEASURE

60 seconds	=	1 minute
60 minutes	=	1 degree
360 degrees	=	1 circle
90 degrees	=	1 right angle
$11\frac{1}{4}$ degrees	=	1 point on the compass

TIME

60 seconds	=	1 minute
60 minutes	=	1 hour
24 hours	=	1 day
365 days	=	1 year

CONVERSION FACTORS _____

Atmosphere (standard)	=	29.92 inches of mercury
Atmosphere (standard)	=	14.7 pounds per square inch
1 horsepower	=	746 watts
1 horsepower	=	33,000 foot-pounds of work per minute
1 British thermal unit	=	778 foot-pounds
1 cubic foot	=	7.48 gallons
1 gallon	=	231 cubic inches
1 cubic foot of fresh water	=	62.5 pounds
1 cubic foot of salt water	=	64 pounds
1 foot of head of water	=	0.434 pounds per square inch
1 inch of head of mercury	=	0.491 pounds per square inch
1 gallon of fresh water	=	8.33 pounds
1 barrel (oil)	=	42 gallons
1 long ton of fresh water	=	36 cubic feet
1 long ton of salt water	=	35 cubic feet
1 ounce (avoirdupois)	=	437.5 grains

THERM-HOUR CONVERSION FACTORS _____

1 therm-hour   =   100,000 Btu per hour.
1 brake horsepower   =   2544 Btu per hour

$$1 \text{ brake horsepower} = \frac{2544}{100,000} = 0.02544 \text{ therm-hour}$$

$$1 \text{ therm-hour} = \frac{100,000}{2544} = \frac{39.3082 \text{ brake horsepower}}{(40 \text{ hp is close enough})}$$

$$1 \text{ therm-hour} = \frac{100,000}{33,475} = \frac{2.9873 \text{ boiler horsepower}}{(3 \text{ hp is close enough})}$$

EXAMPLE: How many therm-hours in a 100-hp engine?
ANSWER: $100 \times 0.02544 = 2.544$ therm-hours

BOILER HORSEPOWER _____

1 boiler hp	=	33,475 Btu per hour
	=	34.5 lb steam per hour at 212°F
	=	139 sq ft EDR (equivalent direct radiation)
1 EDR	=	240 Btu per hour
1 kW	=	3413 Btu per hour

WEIGHT OF WATER AT 62°F _____

1 cu in.	=	0.0361 lb (of water)
1 cu ft	=	62.355 lb
1 gal	=	8.3391 lb (8⅓ close enough)

MISCELLANEOUS DATA

3413 British thermal units (Btu)	=	1 kilowatt-hour (kW-hr)
1,000 watts	=	1 kilowatt (kW)
1.341 horsepower (hp)	=	1 kilowatt
2,545 Btu	=	1 horsepower-hour (hp-hr)
0.746 kilowatt	=	1 hp
1 micron	=	one millionth of a meter (unit of length)

METRIC CONVERSION

NOTE: Now that many countries are using or converting to the metric system, these conversion factors are important to know.

To convert	into	Multiply by
Centimeters	Feet	3.281×10^2
	Inches	0.3937
	Yards	1.094×10^2
Kilograms	Pounds	2.205
Liters	Cubic feet	0.03531
	Cubic inches	61.02
	Gallons (U.S.)	0.2642
	Quarts (U.S.)	1.057
Meters	Feet	3.281
	Inches	39.37
	Miles (land)	6.214×10^4
	Yards	1.094
Myriawatts	Kilowatts	10.0

ARITHMETIC REFRESHER

Handy Multiplication Table

1	2	3	4	5	6	7	8	9	10	11	12
2	4	6	8	10	12	14	16	18	20	22	24
3	6	9	12	15	18	21	24	27	30	33	36
4	8	12	16	20	24	28	32	36	40	44	48
5	10	15	20	25	30	35	40	45	50	55	60
6	12	18	24	30	36	42	48	54	60	66	72
7	14	21	28	35	42	49	56	63	70	77	84
8	16	24	32	40	48	56	64	72	80	88	96
9	18	27	36	45	54	63	72	81	90	99	108
10	20	30	40	50	60	70	80	90	100	110	120
11	22	33	44	55	66	77	88	99	110	121	132
12	24	36	48	60	72	84	96	108	120	132	144

HOW TO USE: To multiply 9 × 9, for example, just read down and over to the center of the table above, and find the answer 81.

Addition: Below is an addition of two numbers as commonly set down:

$$26.13$$
$$\underline{415.38}$$
$$441.51 \quad Ans.$$

Here 3 plus 8 is 11. Put down 1 and carry 1. Then, 1 plus 1 plus 3 is 5, etc. This is the usual procedure and we can't suggest any improvement.

In contrast, try adding a long column of figures. First, the customary way:

$$26.13$$
$$415.38$$
$$260.41$$
$$22.85$$
$$191.06$$
$$205.43$$
$$88.55$$
$$739.11$$
$$\underline{263.05}$$
$$2,211.97 \quad Ans.$$

Following the usual procedure, we added the digits in the last column to get 37. We set down the 7 in the last place of the sum and added the 3 to the sum of the digits in the next to the last column, and so on.

This is faster:

$$26.13$$
$$415.38$$
$$260.41$$
$$22.85$$
$$191.06$$
$$205.43$$
$$88.55$$
$$739.11$$
$$\underline{263.05}$$
$$.37$$
$$2.6$$
$$39.$$
$$37$$
$$\underline{18}$$
$$2,211.97 \quad Ans.$$

Multiplication: RULE: First multiply the two numbers as whole numbers. Then the given product has as many decimal places as are in the *two* numbers combined.

Example: 24.3×0.0613. First, $243 \times 613 = 148{,}959$. Decimal places in two numbers multiplied total five, so correct product is 1.48959.

To check a multiplication, don't repeat the original operation. It is much safer to reverse the numbers and remultiply, like this:

$$462 \times 893 = ?$$

462	Reversing;	839
893		462
1386		1786
4158		5458
3696		3572
412,566 *Ans.*		412,566 *Ans.*

Fractions: To change a proper fraction to a decimal fraction, divide the numerator by the denominator and carry out the quotient to the desired number of decimal places.

EXAMPLE: Solve for decimal fraction equivalent of $7/16$
SOLUTION:

$$
\begin{array}{r}
0.4375 \quad Ans. \\
16\overline{)7.0000} \\
6\,4 \\
\hline
60 \\
48 \\
\hline
120 \\
112 \\
\hline
80 \\
80 \\
\end{array}
$$

EXAMPLE: Solve for decimal fraction equivalent of $5/13$, correct to the third decimal place.
SOLUTION:

$$
\begin{array}{r}
0.384 \quad Ans. \\
13\overline{)5.000} \\
3\,9 \\
\hline
1\,10 \\
1\,04 \\
\hline
60 \\
52 \\
\hline
8 \\
\end{array}
$$

To change a decimal fraction to a proper fraction, multiply the decimal fraction by the denominator of the desired proper fraction and place the product obtained over the denominator.

EXAMPLE: Change 0.032 to 125ths.

SOLUTION: $0.032 \times 125 = 4$
Thus, $0.032 = $ ⁴/₁₂₅   *Ans.*

To change a decimal fraction to the nearest indicated proper fraction, multiply the decimal fraction by the denominator of the desired proper fraction and then round off the product obtained to the nearest whole number and place it over the denominator.

EXAMPLE: Change 0.109 to the nearest 64th.
SOLUTION: $0.109 \times 64 = 6.976 = 7$ *Ans.* (nearest whole number)
Thus, $0.019 = $ ⁷/₆₄   *Ans.* (nearest 64th)

Powers and roots: A power is the product obtained by using a base number a certain number of times as a factor. Thus 3 to the fourth power is written 3^4 and is equal to $3 \times 3 \times 3 \times 3 = 81$ *Ans.*

EXAMPLE: Find the square of 25 (which is written as 25^2).
SOLUTION: $25 \times 25 = 625$ *Ans.*

A root is the opposite of a power. It is one of the equal factors of a number.

EXAMPLE: Find the square root of 64, which is written as $\sqrt{64}$.
SOLUTION: $8 \times 8 = 64$, thus the answer is 8.   *Ans.*
EXAMPLE: Find the square root of 535.713 correct to three decimal places.
SOLUTION: 5'35.71'30 Here we first separate the number into groups of two figures each, starting with the *decimal* point, and going from that point both ways. Next determine the largest number whose square is contained in the left-hand-most group. Thus,

$$
\begin{array}{r}
2 \\
\hline
\sqrt{5'35.71'30} \\
4 \\
\hline
1'35
\end{array}
$$

Now subtract the square of the root ($2 \times 2 = 4$) from the first group and annex the second group (35) to the difference (1'35). A trial divisor is formed by doubling the number (2) contained in the root thus far ($2 \times 2 = 4$). Now determine the number of times this trial divisor is contained in the partial remainder when the right hand digit of the partial remainder is omitted (1'35, omitting the 5 gives us 13). Thus 13 divided by 4 = 3. This same number 3 is also annexed to the trial divisor 4, making it 43. Now multiply the complete divisor by the last number that appears in the root (3), thus $3 \times 43 = 129$. Repeat the last step until the desired number of places have been obtained in the root, thus:

$$
\begin{array}{r}
2\ \ 3.\ 1\ \ 4\ \ 5 \\
\sqrt{5'35.71'30'00} \\
4 \\
\hline
43 \quad 1\ 35 \\
1\ 29 \\
\hline
461 \quad\quad 6\ 71 \\
4\ 61 \\
\hline
4624 \quad 2\ 10\ 30 \\
1\ 84\ 96 \\
\hline
46285 \quad 25\ 34\ 00 \\
23\ 14\ 25 \\
\hline
2\ 19\ 75
\end{array}
$$

Carrying to three decimals gives us 23.145. *Ans.*

NOTE: This has been a brief refresher course, but complete enough to enable the reader to work the problems in this volume. The reader, if deficient, should pursue this subject in a good book on mathematics.

INDEX

INDEX